# breasts

( . ) ( . )

the body literacy library

# breasts

( . ) ( . )

## an owner's guide

## Dr Philippa Kaye

## the body literacy library

Body literacy is a human right. It is a means to observe, learn,
and understand ourselves—three essential steps to enhancing
our self-knowledge and self-care.

With *The Body Literacy Library* you will learn to tune in to every
little bit of yourself. You'll have all your embarrassing questions
answered, and discover everything you need
to know about your body to live a happier, healthier life. This
isn't just about listening to your body, but empowering yourself
with the knowledge of what your body is telling you.

Read this book to love the skin you're in, and make informed,
positive changes to improve your health and well-being—
starting today.

# Contents

(.)(.)

# Introduction

How often do you think or talk about your breasts? Most of us know very little about this part of our body, which is ironic because they have been the subject of debate for millennia, from the ancient stories of one-breasted Amazons to current social media campaigns such as #freethenipple. Yet what people have been talking about is often the objectification of breasts, as opposed to breast health.

Despite all the advancements of the last century or so, women are still judged by how they look. Breasts, and their role in sexuality, arousal, breastfeeding, nurturing, and mothering, continues to be a topic for discussion and opinion. In fact, the very day I write this there is a news report about a woman being told it was "inappropriate" to breastfeed her baby in her car in a supermarket parking lot, while in the supermarket itself there were likely magazines containing sexualized images of women and their breasts.

As a body part, breasts are perhaps only rivaled by genitals for the number of nicknames that they are given: boobies, boobs, bangers, baps, jugs, tits, titties, coconuts, fried eggs, melons, milkers, puppies, the twins, the girls, norks, maracas, funbags, waps, tatas, cha-chas, knockers, rack, shelf, air bags, honkers, babylons, bongos, bazoombas, bazookas, chesticles and more. Few other body parts are given the same amount of thought or airtime. In this book, I'll be using the words breasts and chest interchangeably, no matter the gender of the body being described. Everyone has a right to know their body, and this book is for everyone.

There is an impact of living with breasts, and the current conversation is not about how to look after them. In a study involving 2,000 girls aged 11–18, 87 percent wanted to know more about breast health, showing just how little useful information is out there, and how many want and need this to change. Alarmingly, around half of those surveyed saw their breasts as a barrier to taking part in physical activity (for example, due to breast pain or embarrassment about breasts). Physical activity is essential for health, and has a myriad of well-known benefits, both for physical and psychological health. Anything that can be done to reduce barriers to exercise, including breast health education, must be helpful.

In so many areas of health, the research for the needs of women and girls is lacking, and this includes breast health. The World Health Organization (WHO) definition of health states that: "Health is a state of complete physical, mental and social well-being and not merely the absence of disease or infirmity." This definition applies to breast health, just as it does to other areas, reminding us that it is important to understand your breasts to be able to look after them. For example, why is research on sports bras solely focused on the needs of elite athletes and not recreational athletes or pregnant women?

Women are aware of the specter of breast cancer; we are told of the importance of examining our breasts but often not how to go about it. We discuss making them bigger or smaller, but not why they might be tender and what to do about that. A "wardrobe malfunction" that exposes a nipple to millions of viewers makes the front page of the papers, yet we pay so little attention to our own breast health and care.

This book aims to change that.

# 01

# Breasts are for ...

# what is the purpose of breasts?

The answer is not as simple as you might think—while breasts feed babies and infants, there is so much more to take into account.

For some body parts, the role and function is obvious: your heart pumps blood around your body; your lungs take oxygen out of the air and remove carbon dioxide as waste. But what exactly are breasts for? This is a relatively simple question, but there is almost certainly not an easy answer. The answer is shaped not only by biology but also by society, meaning it may change depending on who is answering.

From a biological point of view, the breasts of humans and other primates are most obviously involved in breastfeeding. Yet we, as humans, differ from other primates: our breasts tend to

stay permanently enlarged from puberty onward. In other primates, breasts only become enlarged when pregnant or breastfeeding, and then shrink down at other times. So why do humans grow breasts during puberty, before pregnancy, and before even starting menstrual cycles? And why do we keep them after our periods stop? There are lots of theories as to why this could be, with hypotheses ranging from the breast's role in sexual attraction to the breast being something for babies to cling onto, even after feeding (after all, a woman's work is never done).

## ANATOMY OF THE BREASTS

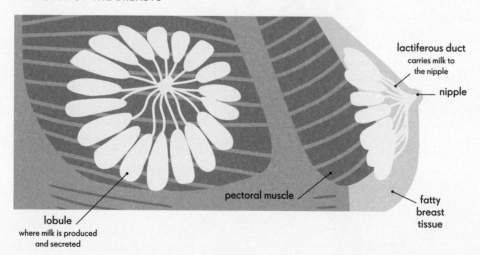

lactiferous duct
carries milk to
the nipple

nipple

pectoral muscle

fatty
breast
tissue

lobule
where milk is produced
and secreted

Breasts and nipples offer more than these biological functions: they are also a potential erogenous zone for all genders, and can play an important role in sexual pleasure and arousal. The hormone oxytocin, known as the love or orgasm hormone, is released during nipple stimulation, which many find pleasurable. This is probably not something you are taught in biology class, but it's important in terms of understanding how we function, and to help us understand our relationship with our breasts.

## The evolution of breasts

Evolution, which has taken place over millennia, has shaped our bodies and their functions into what they are today. Just like every other part of our physiology, breasts play an important role in that story. The term "natural selection" means that those with genes that give a particular advantage are more likely to survive and reproduce. In turn, humans pass those genes on to their offspring. It is likely that there has been an element of selection when it comes to the evolution of breasts.

In the same way that birds develop brightly colored feathers, and chimpanzees and other animals put on courtship displays to attract a mate, perhaps there is a role for breasts in attracting a mate. Although the size of breasts does not indicate the ability to conceive or breastfeed (people are able to lactate irrespective of breast size), breasts may be a basic indication to males that you are a female, so are able to reproduce. Humans, at some point in evolution, stood up, and breasts were more evidently on display (see right).

Have humans changed much from those early primates? One study, featuring men from four countries, showed that irrespective of the size of breasts, respondents reported a preference for firmness. Perhaps this further supports the theory of breasts being a signal to males for reproduction, because although breasts are present after puberty, their size, shape, and firmness change over time (for example, women after menopause, who cannot reproduce, usually have softer breasts).

Another theory relates to the amount of fatty tissue in the breast, which could be seen as a source of energy in times when food is scarce. This is supported by research which shows that someone's physical state may impact breast preferences, with one study showing that men who were hungry preferred larger breasts than those men who had eaten a meal.

### • SENDING A SIGNAL •

When primates such as baboons and chimpanzees are approaching ovulation and are ready to mate, their genital area swells and becomes pinker, probably to signal to potential mates that they are fertile. Humans, at some point in evolution, stood up, meaning that the genitals were not as prominently displayed, so perhaps breasts became a way of showing readiness to reproduce.

## AN IDEAL BREAST SIZE?

A survey of 2,000 found that women and men prefer average-sized (C cup) breasts

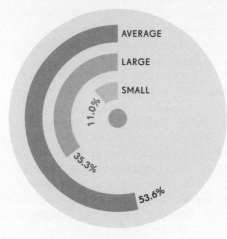

men's preference

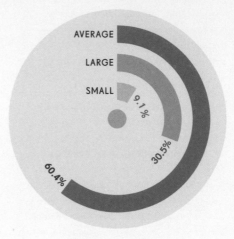

women's preference

## Does size matter?

One study found that women with large breasts and small waists had high levels of estrogen and progesterone. It concluded that this means they are more likely to be able to reproduce. However, this has not been replicated in further studies, and it's important to note that being able to conceive, carry a pregnancy to term, lactate, or breastfeed doesn't relate to your breast size. In terms of fertility and bearing children, breast size does not matter.

Whether or not breast size "matters" will also vary according to global and cultural differences. In fact, recent studies suggest that tastes with regard to sexual attraction and breasts vary enormously among countries, ethnicities and

cultures. For example, men from Brazil reported a preference for larger breasts than those from the Czech Republic, while men from the Azande culture in Africa preferred longer, more pendulous breasts. Men from Papua New Guinea showed a preference for larger breasts when compared to those from New Zealand. Even within the New Zealand group, unmarried men preferred smaller breasts compared to married men. This could be interpreted to support the idea of larger breasts representing increased fertility, but this is contradicted by a further study that reported that men who preferred shorter, non-committed relationships prefer larger breasts.

Research to date seems to be mainly focused on male views and preferences regarding breast size and firmness, with little data about non-cis-heterosexual male's opinions and female opinions in general. Globally, research suggests that heterosexual males prefer medium-size breasts overall. But if being average is preferable for sexual attraction, then why does breast size vary so much?

## Sexual pleasure

Breasts are not solely used for feeding or for the purposes of attracting a mate: they also have a role in sexual pleasure. Most primates do not have sex face to face; in fact, only humans and bonobos do, allowing for easier access and stimulation of the breast area. Nipple stimulation during cuddling and sexual activity releases the hormone oxytocin, which is also released during labor and breastfeeding. Oxytocin is sometimes called the "love hormone" because it makes you feel good and promotes bonding. It could be that nipple stimulation during sex is a way of ensuring humans bond with their partners, because humans are often monogamous.

## Beyond pleasure

None of the discussion so far considers the negative impact of breasts on a person's life. These include unwanted sexual attention, discomfort during growth and the menstrual cycle, and back and neck pain from large breasts (some women choose to have breast reductions in order to manage their symptoms). Breasts can even be fatal, with breast cancer the most common cancer globally. In the UK and US, it is the second most common cause of cancer death in females after lung cancer, and the fourth most common cause of cancer death overall. If there is a purpose for breasts, there is still a relative cost.

Globally, there is a trend toward breast size increasing, and this is not solely related to an increase in breast cosmetic surgery (see chapter 9, pages 184–194). It is likely that this trend is related to diet and lifestyle, as people in general are heavier than previous generations, and their breasts are also getting larger.

Even if there is a biological imperative for large breasts, be that for purposes of sexual attraction, reproduction, or lactation, there will still be an impact on cultural and societal factors which, in turn, may change over time. Is it biology or social training which dictates which breasts are considered most attractive?

Perhaps most interestingly, all the theories mentioned so far focus on the role of the breast for someone *other* than the people whose bodies they are a part of. With regard to milk production, the purpose of the breast is to feed offspring, and with regard to sexual attraction, to attract a mate. Is it any wonder that women may well consider their breasts to be a part of the body that society says never truly belongs to them?

# breasts throughout history

Changing attitudes to breasts reflect changes in society,
from Ancient Greece to the modern day.

The history of our views on breasts can be traced through ancient myths, the arts, and in today's media, encompassing the history of womankind and changing attitudes towards women.

## Mythical breasts

In ancient Greek mythology, the Amazon tribe is heralded as a group of fearsome female warriors, who supposedly removed one of their breasts in order to fire arrows more efficiently. It may be that this nomadic tribe of female warriors really did exist, but the part about removing a single breast (retaining the other to feed only female infants, and discarding male babies), may be the stuff of legend—after all, a bow and arrow can be used without removing a breast. Furthermore, women with one breast are not seen in the artworks of ancient Greece. As for why the tales of a violent tribe of women discarding male babies occurred, perhaps it was a precautionary tale to keep women in their traditional roles?

## Breasts in art

Since ancient times, breasts have been portrayed in art. During the Renaissance period, countless images of the Madonna breastfeeding the baby Jesus were created. These images are thought to symbolize not just feeding Jesus, but the spiritual needs of Christians. At this point in art history, breasts aren't immodest.

One of the most famous paintings, Botticelli's 15th-century *Birth Of Venus*, shows a naked Venus with one breast partially covered. At that time, nudity was used to represent divine love, innocence, or even purity.

The ancient Greek sculpture the *Venus de Milo* (see opposite) is one of the most famous artworks of a woman, and remains a much fetishized icon today. Fast-forward to the 19th century, and women wore corsets to create the curvy body shape of the statue.

Over time, society's views about breasts have shifted. During the 15th century it was documented that King Henry VI felt that baring the breasts was disrespectful, and modest dress became associated with purity and chastity. When Queen Elizabeth I ascended the throne in England, images of her showed the appearance of a flatter chest, perhaps reflecting the reality of a woman in what would have been considered the role of a man.

## Under scrutiny

The bodies of famous women have always been scrutinized. The breasts of Nell Gwynn, the actress and mistress of King Charles II in the 17th century, are well-documented. There was a rumor that the champagne coupe glass was

We look at the debates around breastfeeding on page 18 and in chapter 5.

---

### • THE PERFECT FIGURE? •

The iconic sculpture the *Venus de Milo*, depicting Aphrodite, the goddess of love, shows her with bare breasts. The statue, dating back to around the 2nd century BCE, is thought to have shaped what is considered to be the "ideal" female figure. It has inspired perfume bottles, stage costumes, and much more.

---

Breasts and whether or not they should be exposed continue to be discussed in the media. Remember the furore following the 1999 MTV Video Music Awards when Diana Ross jiggled the breast of L'il Kim, whose nipple was covered by a breast pasty? Or the 2004 Super Bowl, when Justin Timberlake tore Janet Jackson's top and exposed one of her breasts and nipple shield jewelery?

## Double standards

Even now, male and female nipples are not treated equally. In images on social media, female nipples are often blurred or images deleted, and the campaign #freethenipple gains momentum. The Instagram account @genderless-nipples shows only close-ups of nipples, so the gender of their owner cannot be identified.

inspired by the breasts of the famous French Queen, Marie Antoinette (although this is unlikely, as it was actually designed in the previous century). In more recent times, actresses such as Brigitte Bardot and Pamela Anderson have received as much attention for their breasts as for their careers. Somehow, large breasts are now linked with sexual proclivity, promiscuity, and even immorality; consider the press surrounding Anna Nicole Smith (the model who married oil billionaire James Howard Marshall) and the lengthy court disputes over her inheritance.

Breasts and breastfeeding have reflected cultural and societal shifts in history. In Victorian times, breastfeeding was not something routinely done by upper class women. Instead wet nurses were used and only women of what would be considered lower classes would breastfeed their own children. The debate of breast versus bottle is not solely a modern debate, and started in the 19th century, when bottle sterilization practices did not exist in the UK, and women were encouraged to breastfeed as it was associated with lower infant mortality.

Contrast today with 1930s America. Men were not allowed to go topless in many states, even to the beach. Men started protesting and by 1937 the ban was repealed. Meanwhile, women's nipples are still labeled "explicit content" on social media. The female body is constantly deemed as sexual—irrelevant of context—and, in a sense, public property. In other words, a woman is never given full ownership of her body. It is always being observed and judged.

Society seems to have still further issues when breasts are removed. People with mastectomy scars are rarely seen on television, advertising, and social media. It seems that if the breast can't be seen, so, too, the absence of the breast should not be seen.

# breasts and society

The history of breasts and nipples are closely intertwined with cultural and social beliefs around modesty, morality, and sexuality.

Breasts have always been a source of interest and debate, their two "main" roles—the maternal (breastfeeding) and sexual attraction (pleasure)—often held in opposition. This is particularly highlighted when it comes to the topic of breastfeeding in public, with women often being told to cover their breasts and the baby—even though breastfeeding is a natural process and the nipple itself is covered during feeding. In 2012, a *Time* magazine cover showing a woman breastfeeding her three-year-old child made headlines everywhere, leading to debates not just about breastfeeding and attachment parenting, but also about images of breastfeeding in the media.

## Judgments and stereotypes

There are plenty of judgments, assumptions, and stereotypes surrounding the appearance of breasts and their size: large breasts might mark you as "stupid" or a "bimbo," or even give rise to moral judgments about sexual promiscuity. Stereotypes around appearance constantly have an impact on women; for example, some people choose to dress down in the work environment to avoid conflict or attention. Does society allow women to have both breasts and brains? For celebrities like Marilyn Monroe or Dolly Parton, their intelligence is superseded by the mistaken idea that blonde hair and big

breasts mean their owners must not have intellectual value, skills, and talent, and should be admired for their looks alone. This judgment is then balanced against the supposed purity and maternal love around breastfeeding. Contradictory, at best.

What society deems as acceptable with regard to breasts and sexuality can differ depending on where you are. Sunbathing topless is often considered more acceptable in mainland Europe than in America, sometimes called "European sunbathing," yet in some parts of Spain you can be fined for wearing a bikini in the street. The mixed messaging about breasts continues. One study found that women disapprove of female toplessness more than men do—showing again how complicated this issue is.

Girls become aware of their breasts early during puberty, not just due to their own feelings about them, or symptoms such as pain

---

### • BREASTFEEDING TABOO? •

In Sweden, 8 percent of people felt breastfeeding in public was inappropriate in comparison to over 50 percent in Italy. In many countries, women are still expected to "cover up."

or tenderness that they may experience, but also because the time of breast development is often a time when they first become aware of male attention, whether or not it is desired.

If breasts are an everyday feature of the human body, and breastfeeding not shameful, why then hide images of the breast, in particular of the nipple, in the media? And why are we horrified by accidental exposure? Society has sexualized all aspects of the breast, irrespective of how the person with the breasts is actually feeling. Are breasts the ultimate symbol of femininity or under the gaze of others, to be objectified and sexualized?

## Breasts and advertising

Sex sells. Breasts are sexualized, therefore it follows that advertising uses breasts. At the very least, an ad containing breasts gets your attention. Think of the famous ad in 1994 featuring Eva Herzigová for Wonderbra, "Hello Boys." It was reported to be so distracting it caused car crashes (and sold millions of bras). It isn't just ads for bras themselves that use breasts. Breasts are used in advertising to sell everything from food and alcohol to bikes and cars; in fact, the advertising concept behind an entire restaurant chain, Hooters, is based on women and their breasts. Tom Ford advertised its For Men fragrance by showing the bottle between two breasts, and Jean Paul Gaultier's iconic Classique fragrance bottle is in the shape of the female torso in lingerie. There are even companies such as Tittygram, where you can pay for an image of your chosen slogan written on a pair of breasts.

Does it work? It certainly seems to; or at least it gets us talking about the advertisement and the brand. Yet, in 2022, when Adidas launched an ad showing multiple breasts of different sizes, shapes, and skin colors, the Advertising Standards Authority (ASA) in the UK found the poster "likely to cause offense." The ad was also shown on Twitter, and although complaints were made, the company allowed the ad. The statement from Adidas about the ads focused on the fact that breasts are diverse, and therefore their sports bra styles are also diverse, so people can find a bra of the right fit. Although the ASA did not find that the ad was sexually explicit, it found that because it showed explicit nudity it could not be shown untargeted, where it could possibly be seen by children. A counterargument would be that showing real breasts, in all their diversity, could actually help with body image and breast health.

As always, it is complicated, but it seems that it is OK for breasts to be used in advertising to sell lots of things—but only certain breasts, in specific, sexualized ways, as opposed to their diverse reality.

Eva Herzigová's 1994 "Hello Boys" Wonderbra ad was reported to be so distracting it caused car crashes.

# history of the bra

World events have shaped the history of the bra,
just as the bra has shaped the world.

Just as we can trace cultural and sociological views about breasts through the history of art, entertainment, and media, we can also follow the history of the bra and its impact on women and society. Clothes can encompass both fashion and function; a coat keeps you warm but may look stylish, and shoes protect feet while also making a fashion statement. The same applies with bras, which unite function (they support the breasts) and flair, shaping the breasts to give a required aesthetic. World events have shaped the history of the bra, but the bra has also shaped the world—the technology used in the first spacesuits for Neil Armstrong and his colleagues was made by the bra manufacturer, Playtex.

## 4th and 5th centuries BCE

Ancient Greek women in the 4th and 5th centuries BCE were often depicted wearing clothing rather like a bandeau bra, or bands which went over the clothes. By the time of the Romans, bras, known as *"mamillare"* or *"strophium"* were worn under clothes, and were basically pieces of cloth wound tightly around the breasts, and would have given support in the form of compression.

## early bras

The first documented evidence of bras is in the art of the Minoans, a Bronze Age civilization of the Aegean. Images of Minoan women from the 14th century BCE show a linen or leather band, called a *"mastoides,"* which held the breasts from underneath, rather like a shelf or push-up bra, but left the breasts themselves exposed.

## 19th century

In 1869, Herminie Cadolle split the corset in two, with the upper part supporting the bust and the lower part being a corset for the waist, perhaps inventing the bra. Although the combination of bra and corset would still have been restricting, her groundbreaking work was the first step to allowing women a degree more comfort.

At the end of the 19th century, the "rational dress" movement began, and it was felt that corsets posed a risk to health. Tight-fitting corsets were replaced with looser garments and bras. In 1893, Marie Tucek received the patent for the "Tucek breast supporter," which involved two cups and an underband.

## 15th to 18th centuries CE

From the 15th century onward, the corset came into fashion. This garment was generally longer than a bra, reaching down the trunk to both lift the breasts and give shape to the body. Early corsets may have used supports of wood or whalebone, and later used strips of metal. The popularity of the corset has been related to the court of King Henry II of France—apparently his wife, Catherine de' Medici, banned thick waists from court, and the use of corsets to shape the figure became popular. The structure of corsets and girdles changed over the following years to create different figure shapes, according to the fashion at the time.

## 1900s–1910s

The word "brassiere" was first used in the fashion magazine *Vogue* in 1907, and was included in the *Oxford English Dictionary* for the first time in 1911.

The events of World War I accelerated the decline of the corset. The war meant women took on more work and activities which would be impractical in a corset, and all available metal was used in the war effort, leaving none for items like corsets.

The patent for a backless bra may belong to Mary Phelps Jacob (also known as Caresse Crosby), who stitched together two handkerchiefs in 1910 because her corset was visible under her sheer evening dress. She sold the patent to The Warner Brothers Corset Company for $1,500, and they went on to make millions.

## 1920s–1950s

By the 1920s, the fashion for smaller, flatter chests and bandeau-style bras came into being. Then, in the 1930s, the bra as we recognize it developed, and cup sizes were created (see below), followed by underwired bras. The word "brassiere" was now commonly shortened to the word we use now, the bra. As fashions changed, so, too, did bras, with styles that lifted and separated the breasts, padded bras in the 1940s and the bullet- and torpedo-style bras of the 1950s.

### • THE HISTORY OF BRA-SIZING •

It's thought that cup sizes were established in the 1930s, initially by S. H. Camp and Company, who created sizes A-D. Each of the cup sizes developed nicknames: egg cup, tea cup, coffee cup, and challenge cup. It took some time to take off though, and bra sizes were also denoted as small, medium, and large. Underband sizing followed in the 1940s.

( . ) ( . )

## 1960s to 2000s

One of the most famous bra brands, the Wonderbra, was first invented in the 1960s. It lifted the breasts while pushing them together to make the cleavage more prominent, although this bra style didn't really reach mass popularity until the 1990s. The first sports bras appeared in the 1970s, initially made from two jockstraps sewn together. And further changes were made in the 2000s when molded bras began to appear.

## 2020s and beyond

Fashion trends for breast shapes and bras are in constant flux and global events continue to play a role. Research during lockdown in the coronavirus pandemic showed that women moved from buying structured, wired, and padded bras to softer-fitting, non-underwired bralettes and crop-top-style sports bras, perhaps looking for comfort as opposed to appearance. And now, post-pandemic, the trend appears to be reversing.

( . ) ( . )

# 02

# Anatomy and everyday maintenance

# breast anatomy

Let's look at the structure of your breasts, what's
normal, and how you can check your chest.

Did you know that breast tissue extends all the
way from the sternum (breastbone in the center
of your rib cage) to the middle of your armpit?
We tend to think of breasts as the pronounced
circular part of the chest (called the circular
body), but the axillary tail, or "Tail of Spence,"
extends out toward, and into, the armpit.
Breast tissue also extends up the chest and
stops just below the collarbone (clavicle).
This is why when you examine your breasts
(see pages 39-43) you will be checking this
whole area.

## What are breasts made of?

Breasts, or mammary glands, are made of a
combination of glandular and fatty tissues.
Within the glandular tissue there are structures
called lobules, which each contain around 15-20
lobes. This is the tissue that produces milk, if
needed. Each lobule connects to milk ducts,
which then connect to the nipple. The lobules
themselves are clustered, rather like a bunch of
grapes. In between these lobules are fatty and
connective tissues.

The breasts also contain blood and lymph
vessels as well as nerves for sensation. Breasts
don't contain any muscle, but sit on top of the
muscles of the chest wall (the pectorals). They
do contain ligaments, such as Cooper's
ligaments, and other connective tissue for
support and shape. The size of the breast is
mainly determined by the amount of fatty
tissue within it.

The male breast region is extremely similar,
and men do have some breast tissue, but
usually a smaller amount.

(.)(.)

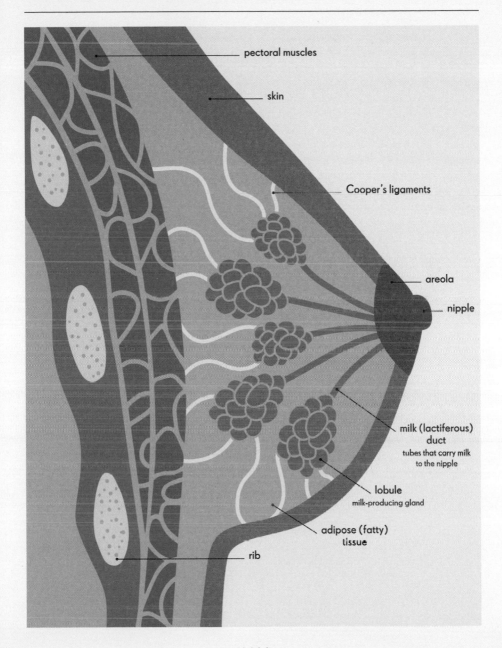

pectoral muscles

skin

Cooper's ligaments

areola

nipple

milk (lactiferous) duct
tubes that carry milk to the nipple

lobule
milk-producing gland

adipose (fatty) tissue

rib

( . ) ( . )

## The lymphatic system

Lymphatic vessels carry lymphatic fluid around the body and drain to lymph nodes. The armpits contain around 20–30 lymph nodes, and the lymphatic vessels in the breast drain around 75–90 percent of fluid to the lymph nodes in the armpit—the rest mainly drain to the nodes around the breastbone. The lymphatic system has multiple roles, including:

• **Acting as a drainage system** by removing excess lymphatic fluid from body tissues and draining it into the bloodstream.

• **Fighting infection**—the lymph nodes contain lymphocytes (white blood cells), which deal with viruses, bacteria, and damaged cells.

• **Removing waste** made by the cells (see chapter 8 for more on lymph nodes and breast cancer).

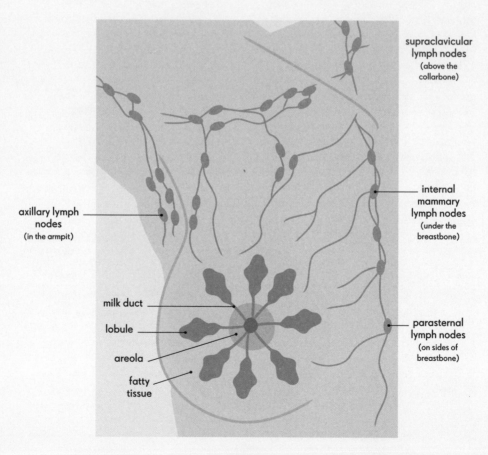

supraclavicular
lymph nodes
(above the
collarbone)

internal
mammary
lymph nodes
(under the
breastbone)

axillary lymph
nodes
(in the armpit)

parasternal
lymph nodes
(on sides of
breastbone)

milk duct

lobule

areola

fatty
tissue

(.)(.)

# breast development

Breasts begin to develop in the womb and continue to change and evolve throughout our lives.

Breast formation begins—perhaps before many people realize—when the fetus is developing in the uterus, with a thickening called the mammary ridge. Breasts and nipples begin to form in the very early stages of fetal development (at about 4-6 weeks after conception). By the time the Y chromosome kicks into action, and testes begin to develop (at about 7-9 weeks of pregnancy), the nipples are already in place, which is why all genders have breasts and nipples.

By the time of birth, a baby's milk duct system and nipples have developed. The breasts of males and females remain relatively identical until puberty, when the mammary glands in female breasts develop further. In most cases, the male breast remains undeveloped, unless gynecomastia occurs (see page 135).

## Early breast development

The majority of newborn babies (between 65-90 percent) will have some breast tissue due to the transfer of the hormones estrogen and progesterone from the mother during pregnancy. Newborn breast buds are common and can occur in all sexes, appearing in the first week or so after birth, most often on both sides of the body. As with adult breasts, they may not be symmetrical. Breast buds appear like small disk-shaped lumps under the chest area. Usually, no treatment is required as the enlargement tends to go away on its own within a few weeks. The swelling may last longer in breast-fed babies, up to about six months. While you are waiting for the swelling to recede, do not stimulate the breast tissue, or try to squeeze out milk, as this may irritate the tissue, worsen the enlargement, or lead to an infection.

### • LEAKING NIPPLES IN BABIES •

Changes in hormone levels can stimulate the baby's brain to produce the hormone prolactin, and produce a clear or milky-colored fluid from the nipple. This generally goes away on its own.

## WHEN TO SEE A DOCTOR

If breast development is asymmetric—occurring only on one side and not the other—or if the skin over the breast is red, warm to the touch, or swollen/hard, or if your baby is unwell, then other diagnoses such as infection may need to be considered. If your baby still has breast buds at six months of age, be sure to discuss this with your doctor.

## WHY DOES MY TODDLER OR PRESCHOOLER HAVE BREASTS?

Breast development tends to start at around 8 years old, and is considered premature if changes start before this. "Premature thelarche," as it's known, can affect one or both sides of the body. It can occur on its own, where the cause may not be known, or be associated with premature puberty.

In premature thelarche without premature puberty, the breast development tends to be small (generally not past Tanner stage 3, see page 47). On its own, premature thelarche does not usually require further investigation or treatment, and in many cases, the breast development may regress spontaneously or stop. Your doctor may monitor to check for other signs of puberty, such as growth of pubic hair or rapid physical growth, and may request X-rays to check bone age. In premature thelarche without premature puberty, puberty usually starts later, at around age 11.

### PREMATURE PUBERTY

After the age of three, premature thelarche is more likely to be associated with premature puberty (puberty that starts before the age of 8). If early puberty is a possibility, then it is likely you will be referred to a pediatrician or pediatric endocrinologist for investigations to rule out an underlying cause that needs treatment, and to consider medications to delay puberty.

## Life changes

As a female child reaches puberty, the ovaries begin to produce estrogen. This triggers the development of the breasts. Small divisions of breast tissue (lobes) develop, followed by mammary glands (which consist of 15–24 lobes).

From the age of around 35, the milk ducts and mammary glands begin to shrink (involute) as estrogen levels begin to fall when females enter perimenopause and after menopause, the changing hormones may cause the breasts to change size and shape.

Breasts may also change size and shape if you lose or gain weight. You may notice that the breasts increase and decrease in size significantly over the menstrual cycle; you may even measure different bra cup sizes at different times of the month. Other events, such as surgery or radiotherapy, may also change the appearance of the breast.

**Premature thelarche is the term for premature breast development, when the breasts start to develop earlier than expected.**

# are my breasts normal?

You are unique, and so are your breasts. In this section, we look at breast size and shape, nipples, and areolas, and work out what is normal for you.

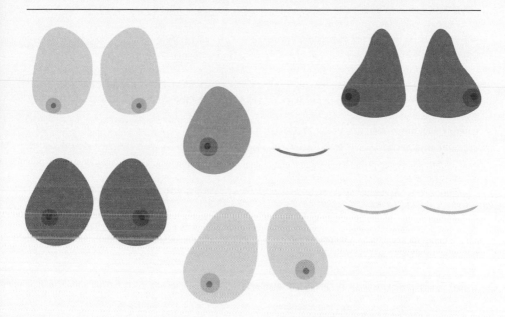

## Size and shape

Everyone's breasts are different in size and shape, and it is normal for there to be some differences between your breasts.

- **Asymmetry is completely normal** and very common. Some bras can change your breasts' appearance or may have padding, so different styles may cover some degree of asymmetry.

- **One breast may be bigger or smaller** than the other up to about a cup size, or the breast or nipple may sit slightly higher or lower than the other. You may notice that the breasts have a different shape to each other.

- **Your breasts are likely to change** in size and shape at different stages of your life; for example, if you become pregnant or are breastfeeding.

(.)(.)

## Nipples

- **Nipples vary** in color, size, and shape. They may be in different positions; point up, down, or in different directions; and they may be slightly different on each breast.

- **Most nipples stick out** a few millimeters above the areola. This may be more prominent in some people, and can become more obvious if you are cold or aroused. But they can lie flat, too.

- **Nipples can be inverted,** and about 10 percent of us have at least one inverted nipple, where the nipple points inward. Depending on the grade of inversion, they may pop out when cold or aroused. And yes, it is possible to breastfeed with inverted nipples.

- **Some nipples are cone-shaped,** blending in with the areola. This is a normal part of development, but in about a third of people it stays this way after puberty, and may change again after pregnancy.

- **If your nipples change** in any way from their normal appearance, see your doctor (see pages 40-43).

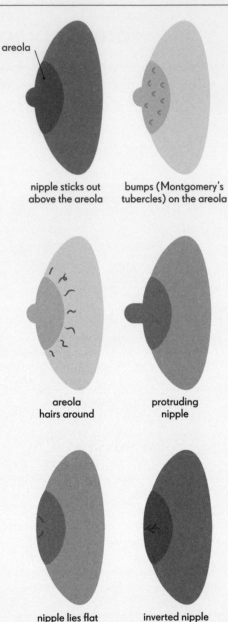

areola

nipple sticks out
above the areola

bumps (Montgomery's
tubercles) on the areola

areola
hairs around

protruding
nipple

nipple lies flat

inverted nipple

# Areola

- **The areola is the darker-colored skin** around the nipple. Just like nipples, and breasts themselves, areolas can be different colors, shapes, and sizes. They may also be different on each breast.

- **It is common to have hair** around the areolas, as the skin there contains hair follicles. This might mean that you have tiny, fine, pale hairs or darker, thicker hair. Both are normal.

- **Bumps are normal, too.** These are called Montgomery's tubercles and may be present on one or both areolas. Sometimes they may look like tiny white spots. These help secrete oils to keep the skin healthy and protect it during breastfeeding.

- **Areolas have nerve endings**, as do nipples, so their stimulation can help with arousal. Stimulation of the areolas during breastfeeding triggers the release of the hormones oxytocin (see page 95) and prolactin to produce and release milk.

# Stretch marks

- **Stretch marks can appear on the breasts** as well as other parts of the body. They tend to occur during breast development as the breasts grow quickly, but they can occur at other times, such as during pregnancy or if your weight changes.

- **Stretch marks are caused by the skin stretching quickly,** which can cause the collagen and elastin in the skin—the proteins that help support the skin—to rupture. A stretch mark is essentially a form of scar.

- **Not everyone will develop stretch marks,** although they seem to be more likely if other people in your family have them. The fluctuating hormone levels of puberty may also be a factor.

- **Stretch marks tend to appear as narrow lines** that start out as red, purple, pink, brown, or reddish-brown in color, but fade over time. They may also start off slightly raised, and over time sink down into the skin, causing a slight dip in the surface of the skin. Your stretch marks may be more obvious if your skin becomes tanned, as the mark itself does not tan.

- **There is no universally effective treatment** for stretch marks. There is some evidence that products containing hyaluronic acid may be helpful at treating and preventing stretch marks. It may be that massaging a product into early stretch marks is more effective than trying to treat old ones.

## Lumps and bumps

- **Breasts may feel lumpy or smooth,** and both might be normal for you. In fact, if it is also normal that your breasts change in texture throughout your menstrual cycle, you may notice that they become more lumpy in the second half of your cycle—in the lead-up to, and during, your period—and then change again afterward.

- **You may notice that** your breasts are always lumpy, almost as if you have frozen peas under the surface. These are known as "nodular" breasts, and if they are on both sides and don't change, this may be normal for you. If anything changes in your breasts, please see your doctor.

## Extra nipples

- **Extra nipples occur in 1 to 5 percent of people.** Also known as supernumerary, accessory, polythelia, or vestigial nipples, an extra nipple may be so small you don't even realize you have one, or presume it is a mole or chickenpox scar.

- **Extra nipples tend to occur along the milk line,** where the breast tissue forms in the womb, between the armpits and the groin.

- **They can occur on one side or both**; may be attached to some underlying breast tissue, or not; and may sometimes have an areola.

- **Some people may notice changes** to their extra nipple during puberty, as hormones fluctuate during the menstrual cycle or during pregnancy or breastfeeding.

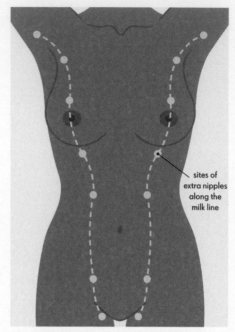

sites of extra nipples along the milk line

**MILK LINES**

---

### • ACCESSORY BREASTS •

Just as you can have extra nipples, you can also have an extra, or "accessory" breast or breasts. These may be apparent from birth, or you may only notice them during puberty. They tend to be in the lower armpit and may or may not have nipples. Having extra breasts or nipples is not medically concerning, but be sure to examine them when you do your self-checks.

---

( . ) ( . )

# the nipple

The nipple is the center of the areola and may sit high or low on
the breast. Let's look at the nipple in detail.

## ANATOMY OF THE NIPPLE

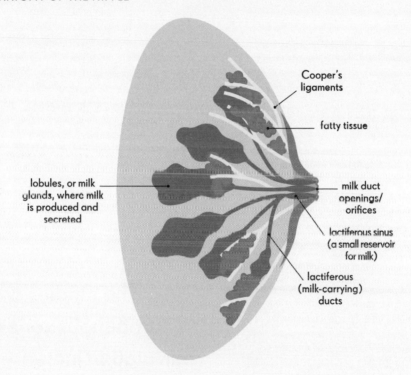

Cooper's
ligaments

fatty tissue

lobules, or milk
glands, where milk
is produced and
secreted

milk duct
openings/
orifices

lactiferous sinus
(a small reservoir
for milk)

lactiferous
(milk-carrying)
ducts

Each nipple contains about 5-20 small holes, called milk duct orifices, which connect to the milk ducts. If you breastfeed, this means there are multiple small streams of breast milk.

Nipples also contain many nerve endings, making them extremely sensitive to touch. See pages 32-34 for more on the development of nipples, variations, and extra nipples.

(.)(.)

## Why do we have nipples?

Nipples exist not only to deliver breast milk, but also for sexual pleasure and satisfaction. The nipples are erogenous zones for all genders, and stimulation increases sexual satisfaction and releases oxytocin, which is involved in orgasm. In fact, stimulation of the nipple can even lead to a "nipple orgasm," where orgasm is reached by stimulating the nipples and breasts alone. This can happen because stimulation of the nipples triggers activity in the same region of the brain as when the genitals are stimulated.

## Why do nipples become erect?

The nipples are full of nerve endings, and can respond to stimulation, be it physical, such as a change in temperature, or psychological, such as stress or arousal. Touch, in all forms, can lead to the nipples becoming harder; for example, if clothing brushes against the nipple.

Cold is often a trigger for the nipples to become erect, as when we are cold the muscles under the skin contract to trap warm air near the skin, causing goosebumps. The smooth muscle under the areola contracts, pulling the skin, causing the nipple to become erect. This happens involuntary and isn't something we can control.

Nipples also become erect during sexual arousal; here, the hormone oxytocin is released, causing the smooth muscle under the skin to contract and the nipple to become erect. Your nipples may also respond to the hormones of your menstrual cycle; for example, you may notice that your nipples are more prone to becoming erect when you are ovulating, during your period, or when you are pregnant.

Nipples may become more or less sensitive after nipple piercing, so may be more or less likely to become erect.

The nipples are erogenous zones. Stimulation increases sexual arousal and releases the hormone oxytocin.

( . ) ( . )

## Inverted nipples

Although estimates vary, it is thought that around 10 percent of us have inverted or retracted nipples to some degree, and this can affect both males and females. Retracted nipples may have the appearance of "slits" within the areola, and this is a normal variant of the nipple.

## Causes of nipple inversion

Nipple inversion is often congenital, i.e. something we are born with, either due to the base of the nipple staying small during breast development in the womb, or the milk ducts forming differently, leading to the nipple being pulled inward. There are other, non-congenital reasons, too, including:

• **Menopause**—as females age and experience the hormonal changes of perimenopause and menopause, the milk ducts of the breast shrink and shorten, which may pull the nipple inward.

• **Mammary duct ectasia**, a condition where the milk ducts widen, can also lead to nipple inversion.

• **Trauma** to the milk ducts, which might occur during breastfeeding, injury, or breast surgery, can cause the nipple to invert.

• **Mastitis**, or breast infection, which is common during breastfeeding, can lead to nipple inversion and is generally treated with antibiotics (see page 110).

## GRADES OF NIPPLE INVERSION

**GRADE 1** the nipple looks inverted but may protrude outward upon stimulation or when it is cold.

**GRADE 2** the nipple looks inverted and although it can be pulled out, it reverts back inward right away.

**GRADE 3** the nipple cannot be pulled outward.

(.)(.)

- **An abscess,** an area filled with pus, can form under the areolas and cause nipple retraction. An abscess may be associated with mastitis, or occur related to a nipple piercing or diabetes. It is generally treated by drainage and antibiotics.

- **Breast cancer** can cause nipple inversion or flattening (see page 160).

## TREATMENT

Although benign flat or inverted nipples do not require treatment, some women may wish to correct the inversion due to cosmetic or functional concerns. Treatment isn't medically necessary and will depend on the grade of nipple inversion. If you have always had flat or inverted nipples, then they may be your normal, but if you notice a change to the nipples, whether on one side or both (see page 40), then please see your doctor for further assessment.

There are certain exercises that can help the nipple stand out. These are often used in preparation for breastfeeding:

**01.** Place your thumbs on either side of the nipple and push downward, while gently, but firmly pulling your thumbs apart.

**02.** Work your way around the nipple repeating the technique, and the nipple may start to protrude outward.

Nipple eversion devices in the form of suction cups, syringes, or breast shells are available, which apply pressure around the nipple, encouraging it to pop out. Alternatively, surgery may be an option, although this isn't always successful, with some nipples inverting again afterward. Surgery may cause damage to the milk ducts, which could then affect the ability to breastfeed in the future, so please discuss with your doctor (see page 102.) Piercing the nipple might also be useful (see page 84).

---

### • BREAST-SCREENING •

Breast-screening is an important part of routine breast care and maintenance. Approximately 1 in 8 women will develop breast cancer, and screening aims to pick this up at an earlier stage than it might otherwise present; for example, before you might notice a lump. This is because, with cancer, time matters; the earlier that a cancer is found, the more likely it is that you will be able to have treatment and recover.
For more information on breast-screening, see page 119.

# being breast aware

Regularly examining your breasts will help you know
what is normal for you, and spot any issues.

Knowing what is normal for you is key—if
you know what your breasts normally look and
feel like, then you are more likely to notice if
something changes, so you can get it checked
out with your doctor. If you do find a change
in your breasts, and if it turns out to be breast
cancer, the earlier the diagnosis, the more likely
it is that it will be easier to treat and that you will
make a full recovery.

Everybody, whatever their biological sex or
gender, should check their breast or chest area,
generally once a month. You may notice that
your breasts change throughout your menstrual
cycle, if you have one, so it is useful to assess
at different points during the cycle so you learn
what is normal for you at different stages.

You may choose to examine your breasts just
after your period as your breasts are less likely
to be tender or lumpy at that point. If you don't
have periods—for example, if you have been
through menopause or are male—then a good
reminder is to examine on the first day of
each calendar month: Feel on the First!

Nearly 8 out of 10 women in the UK aren't
checking their breasts each month, and around
a third never check them at all. Reasons
given include: forgetfulness (35 percent);
time restraints (22 percent); lacking confidence
to self-assess (17 percent); and fear of what
they may find on self-examination (13 percent).

Age and breast size also play a role in checking
rates. Half of women aged under 40 thought
breast cancer only affected women over the
age of 50. Just under half knew that it could
also affect those with smaller breasts.

There are differences in checking between
ethnic groups, with rates lowest among South
Asian women and among Black women.
Reasons given include stigma around
talking about breast health (12 percent) and
embarrassment about checking (10 percent).
Education within families is also important, with
around a quarter of Black women saying that
they weren't taught about breast checking by
relatives, and about one in six reported not
feeling comfortable about self-checking.
Representation matters: over a quarter of Black
and South Asian women reported that they
don't see representation of their groups in
awareness campaigns.

## WHAT TO LOOK OUT FOR

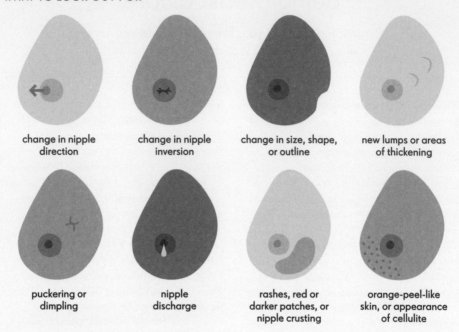

change in nipple
direction

change in nipple
inversion

change in size, shape,
or outline

new lumps or areas
of thickening

puckering or
dimpling

nipple
discharge

rashes, red or
darker patches, or
nipple crusting

orange-peel-like
skin, or appearance
of cellulite

## HOW TO EXAMINE YOUR BREASTS:

### Look

**01.** Stand in front of a mirror, with your hands relaxed by your sides, and look at your breasts. Turn to one side and then the other. If you have large breasts, lift them up so you can look underneath.

**02.** Look for any changes to your breast size and shape, or any new asymmetry. Look for any lumps, or changes in your breast outline.

**03.** Look for any changes to the nipples. Is either nipple pointing in a different direction than

usual, or become inverted? (Remember, only a change to the nipples needs assessment, not if you have always had inverted nipples).

**04.** Look for any skin changes: Can you see a rash; red or pink patch; wet-looking patch, crusting, or scaly looking skin? Is there any itching on the breast, areola, or nipple? Is there any dimpling, like the surface of an orange (called "peau d'orange").

**05.** Is there any nipple discharge (if you aren't breastfeeding)? It could be clear, milky,

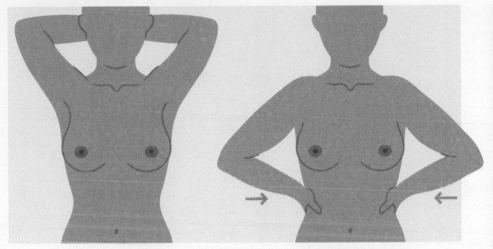

raise your arms and take
another look at your chest

with your hands on your hips,
push inward to tense your chest,
and look at your breasts again
from all sides

or bloody.

**06.** Look for any puckering or dimpling of the breast; this can look like an area is tethered to something underneath, causing the breast to wrinkle, dimple, or pucker.

**07.** Now raise your hands above or behind your head (like you are sunbathing) and take another look. Place your hands on your hips and push inward as this tenses the chest muscles and will make tethering or dimpling more obvious, and check again.

(.)(.)

## HOW TO CHECK YOUR BREASTS

examine the whole breast
up to the collarbone and
into the armpit

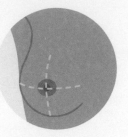

you can examine each
quadrant in turn

or feel in a circular motion
starting from the nipple and
working outward (or vice versa)

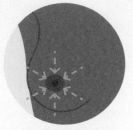

or, examine the breast in
wedge-shaped sections

look out for any pain
or discomfort

## HOW TO EXAMINE YOUR BREASTS:

### Touch

You can check in any position that you find
comfortable: standing, sitting, lying down,
whatever works for you. It may be easiest to
examine your breasts in the bath or shower
using a wet, soapy hand.

Lying on your back slightly propped up on a
pillow can help bring the breast tissue up onto
the chest so you can examine it. If your breasts
are large and flop outward toward your sides,
you can roll slightly toward the opposite side

of the breast you are checking to bring the
breast tissue onto the chest so you can examine
it properly.

**01.** Lift one hand above your head and use the
other hand to examine the opposite breast.
The breast tissue extends up to the
collarbone and into the armpit, so it is
important to examine the whole area.

**02.** In order to ensure that you have examined
the whole breast area, imagine dividing the

breast into quarters and examine each in turn. Or imagine that each breast is a clock and examine the wedge-shaped areas between the numbers, moving toward the nipple. Or you could move your hand in a circular motion, starting on the outside of the breast and making smaller and smaller circles until you cover the whole breast and nipple. Or, start from the inside and work your way out. It doesn't matter how you do it, just that you do it! Ensure you examine the nipple and all the way up to the collarbone as breast tissue can extend that far up. Then examine up to, and in, the armpit as the breast tissue extends to the armpits which also contains lymph nodes.

03. Use the flat of your fingers to feel the breast, searching for any changes. As you press with your hand, you are pressing the breast tissue against your chest wall so you can feel if there are any changes underneath. It will take some time to get to know your own breasts; some people describe that they feel like blancmange or jelly. There may be small lumpy areas that feel rather like a collection of frozen peas, especially around the time of your period.

04. Changes could be a lump, a thickening of the breast, or a bumpy area. The upper outer section of the breast is where you are most likely to find small lumps called nodularity. If normal to you, nodularity is usually felt in both breasts. If it is felt in one breast alone, or is a change for you, it should be checked by your doctor.

05. Any discomfort or tenderness should also be assessed, especially if it is new or only affects one breast (pain is a rarer symptom of breast cancer than other symptoms).

## How to examine your breasts after breast-cancer surgery

It's important to get to know what is normal for your chest after surgeries like lumpectomy or mastectomy. Local recurrence of breast cancer can happen, depending on the type of breast cancer you have. This may range from about 2 to 12 in every 100 breast cancer patients. Examine your chest as described in this chapter, first looking in the mirror and then feeling with your hand. In particular, look and feel for any lumps close to your scar. Are there any skin changes like a small ulcer or a rash? Your scar may extend to your side so be sure to check the whole area.

---

**• THE TLC OF BREAST EXAMINATION •**

Touch
Look
Check changes with a doctor

If you notice any changes to your breasts, get checked by your doctor. This doesn't necessarily mean that you have cancer— it simply means that you have noticed a change that needs to be checked.

---

# 03

# During puberty

# breast changes during puberty

Puberty is a time of great change in both mind and body.
Let's look at breast development, and choosing
the right bra for you (if you want to wear one).

You might be reading this going through puberty yourself, or as a parent or carer of a young person. It is important to remember that everyone is unique and we all grow and change at different times. For some, puberty can be an awkward and uncomfortable time, but understanding what is happening to your body can help. Wearing comfortable clothes, perhaps a bra if you want, is part of this self-care.

Puberty can start from around 8 years old (see page 30) although this varies due to factors such as your genes and weight. During puberty, a part of the brain called the hypothalamus starts to release gonadotrophin releasing hormone (GnRH), which stimulates another part of the brain, the pituitary gland, to produce follicle stimulating hormone (FSH) and luteinizing hormone (LH). These hormones work in combination to stimulate the ovaries to start producing estrogen and progesterone, which have effects all over the body, as anyone who has been a teenager (or lived with one) can testify. The breasts begin to grow and the duct system inside them starts to form. Over time, as the menstrual cycle starts, the breasts further mature, glands form at the end of the milk ducts, and the lobules develop.

All this change may mean you experience tenderness, aching, itching, or tingling in your chest. Nipples can feel swollen and tender.

Breasts can develop at different speeds, so you may notice an aching sensation in one, or a change in size in one and then the other—this is all very common and normal.

As the menstrual cycle develops, your breasts may change throughout your cycle, and can feel very different on different days of your period. It can be worth tracking your cycle to notice any patterns.

## • DELAYED DEVELOPMENT •

If you have not started to develop breasts by the age of 13 (stage 1-2, see opposite), or you develop breasts but have not started your periods by 15, please see your doctor. There may be an underlying cause of "delayed puberty," such as illness, a hormonal or genetic condition, or malnutrition. If your doctor is considering delayed puberty, they may refer you to see a specialist for further evaluation and treatment, if needed.

## The stages of breast development

There are five general stages of breast development (thelarche), which doctors categorize using the Tanner Classification system. Stage 1 tends to start between 8-11 years of age.

The breasts after adolescence are dense, because they contain the milk systems in case pregnancy occurs. However, as you move toward and beyond menopause, the lower hormone levels mean that the glandular tissues shrink, leaving the breasts feeling softer because they are mostly fatty tissue.

The size of your breasts is due to a combination of genetic and environmental factors and is mostly related to the amount of fatty tissue within the breast.

**STAGE 1** Prepubertal, only the nipple is raised.

**STAGE 2** Breast buds (hard, disk-shaped lumps under the nipple) develop, causing the breast and nipple to be raised. The areola increases in size.

**STAGE 3** Breasts continue to grow and develop glandular breast tissue.

**STAGE 4** The areola and nipple become raised, looking slightly higher, or separate, from the rest of the breast.

**STAGE 5** Mature teenage or adult breasts.

(.)(.)

# is something bothering you?

It is normal for your breasts to vary in appearance. Let's look at why some people might talk to a doctor about how their breasts look.

Breasts change throughout your lifetime. However, if the shape, size, or symmetry is worrying you, or seems extreme, do talk to your doctor.

## Amastia, amazia and athelia

If breast tissue, nipples, and areolas do not form at all, you might have a rare condition known as amastia. If the nipples and areolas are present but there is no breast development, that is known as amazia. And in athelia, the breast develops without a nipple or areola.

## Breast hypoplasia

There is breast tissue, nipples, and areolas but they do not develop during puberty or may develop only slightly.

## Poland Syndrome

A rare condition also known as Poland's syndactyly. Here, the pectoral chest muscles don't develop and there is no breast or nipple development. The underlying ribs and sternum can also be affected.

## Tubular or tuberous breasts

This occurs when the breast tissue doesn't develop properly, leading to one or both of the breasts becoming tube-shaped. There may be a large gap between the breasts and they may have a very large or prominent areola and nipple.

### TREATMENT

Problems with underdevelopment of breasts can be treated surgically, often with the insertion of a tissue expander implant, which is then exchanged for an implant once the desired shape and size has been developed. Nipples can be constructed surgically, and tattooing used to give the appearance of an areola and nipple.

## Large breasts

Breast hypertrophy is the medical term describing breasts that grow disproportionately large for the frame of the person. Breast hypertrophy can occur from puberty, or later in life, sometimes in relation to pregnancy or weight gain. It can also be related to certain medications. If it is related to pregnancy, then the breasts may return to their pre-pregnancy size after birth, or at the end of breastfeeding.

Sometimes breast hypertrophy is described using the terms *macromastia* and *gigantomastia*, depending on the size of the breasts. The diagnosis and management are focused on your symptoms and the impact they are having on your life.

Large breasts are heavy, meaning that you are carrying around two extra weights on your chest all day. This can affect your posture, and cause or contribute to pain in your neck, shoulders, and upper back; headaches; or numbness or tingling in the arms. The weight of the breasts may mean that they hang or sag down and rashes, fungal infections, or ulcers can occur in the skin folds underneath the breasts (see page 133). There can be pain in the breasts, or tingling or numbness in the nipple. There might also be pain and discomfort caused by bra straps digging into your shoulders. Large breasts can impact your ability to exercise in comfort.

Of course, breast hypertrophy may not only affect physical health; it can have an impact on mental health, too. It can be difficult finding undergarments and clothes that fit over the breasts, which can lead to body image issues and low self-esteem. If you do find supportive bras and well-fitting clothes they may be more expensive.

You may feel uncomfortable or embarrassed thanks to unwanted attention or comments, which can impact your mental health. Don't underestimate the effect that your breasts may have on your well-being; if they cause you pain or distress, be sure to get the help you need.

### TREATMENT

If the size of your breasts is causing you significant issues, and conservative measures such as wearing supportive bras aren't helping, you may speak with your doctor about breast reduction surgery. Reduction mammoplasty can change the cosmetic appearance of the breasts, but is commonly performed to address symptoms of chronic pain and overall quality of life (see page 186). It's worth noting that surgery may impact your ability to breastfeed, and breasts can increase in size again during pregnancy or if you gain weight.

( . ) ( . )

# breast development

## CAN I SPEED UP BREAST DEVELOPMENT?

No, breast development cannot be accelerated. However, hormonal treatment can encourage the development of breast tissue in transgender women.

———

## WILL MASSAGE HELP WITH GROWTH?

Breast massage won't affect the size of your breasts, and if your breasts are tender, massage may be painful or uncomfortable.

———

## ARE THERE CREAMS OR SUPPLEMENTS TO MAKE MY BREASTS BIGGER?

There is no evidence that creams or supplements marketed at increasing your breast size are effective.

———

## DOES MY WEIGHT AFFECT THE SIZE OF MY BREASTS?

Sometimes, but not always! There is fatty tissue in the breasts, so gaining and losing weight can change the appearance of your breasts. If you lose weight, your breasts might appear smaller. If you put on weight, they might appear larger, but there are other contributing factors, such as genetics.

———

## WILL CONTRACEPTION MAKE MY BREASTS BIGGER?

Hormonal forms of contraception may temporarily increase the size of your breasts and may also contribute to breast tenderness. The change in size isn't permanent. For more information on contraception and breasts see pages 73 and 151.

———

## WILL EXERCISE CHANGE MY BREASTS?

Your breasts don't contain muscle so exercise won't affect their development. However, your breasts sit on top of the pectoral muscles, so exercising these muscles may affect the appearance of the breasts.

———

## IF I WEAR A BRA THAT IS TOO TIGHT OR TOO LOOSE, WILL IT AFFECT MY BREASTS?

No, wearing a bra that does not fit properly does not affect your breast development. But it might be uncomfortable. See pages 52-67 for more on bras.

———

## SHOULD I SLEEP IN MY BRA?

That is up to you. If you find it more comfortable to sleep in a bra then that is your choice—it won't affect how your breasts develop. If you have heavy or large breasts, sleeping in a bra can help with breast discomfort at night.

———

## DOES SLEEPING ON MY FRONT DAMAGE MY BREASTS?

No, how you sleep doesn't affect the development of your breasts, but it may be uncomfortable if your breasts are tender.

———

## MY FRIEND HAS STARTED DEVELOPING BREASTS EARLIER THAN ME. WILL HER BREASTS BE BIGGER THAN MINE?

The time and speed of breast development doesn't affect the eventual size of the breasts, so if someone starts developing earlier or later than you, it doesn't mean that they will be bigger or smaller in the end.

———

# bra basics

Choosing a bra for the first time (and later!) can be overwhelming. There is so much choice in sizing, style, fabric, appearance, and price—so where do you start?

Every day in my office, I see bras that do not fit properly: breasts may not fit in the cup, the band may be riding high between the shoulder blades, or the shoulder straps may be falling off. It is often quoted that around 80 percent of women are not wearing a correctly fitting bra.

In 2005, Oprah Winfrey hosted a segment on her show called The Bra Revolution, where she encouraged women to get fitted, saying that the right size bra "literally performs miracles."

## What about a first bra?

Everyone's breasts develop at a different rate. During puberty you can experience periods of rapid breast growth, meaning that new bras may be required every few months, so affordability may be an important factor when you are choosing a bra. Styles with soft fabrics and without seams might be more comfortable, especially as you get used to wearing a bra. Bras with adjustable fastenings and straps are very important for a comfortable fit. You don't need to avoid underwired bras, as long as the bra fits (see page 63).

Although buying online may be convenient, bra sizing and fitting isn't easy, so if you can, go to a store with dressing rooms so you can try on a range of styles and sizes, and with a fitter, if possible, so you can get good advice.

---

### Q: DO I NEED TO WEAR A BRA?

——————————————— ( . ) ( . ) ———————————————

**A:** *Up to you! But if you are going to wear a bra it seems sensible to wear one that fits. Wearing a bra helps support the breasts and can help avoid breast discomfort, especially when exercising. Breast sagging is related to multiple factors but we know that in cultures where bras are not traditionally worn, there is significant ptosis (sagging), so you might choose to wear a bra for this reason. Wearing a well-fitted bra, even if worn during the day and at night, does not cause health problems.*

( . ) ( . )

## How often should I get measured?

You should see a bra fitter, or measure yourself, if there has been any significant change in your life; for example, if you have grown, gained or lost weight, or if you have been pregnant or breastfeeding. You may notice that your breasts change significantly during your menstrual cycle and that you need different bras at different times of the month. Even if you think you know your size it can be useful to measure again approximately once or twice a year as your breasts change. If your bras feel uncomfortable, then you can check more frequently.

## When should I get a new bra?

Even if you haven't changed bra size, bras eventually wear out. The band of a bra will stretch over time, though they are designed to compensate for this with multiple hook-and-eye settings, so you can use a tighter setting as time passes. However, if the bra feels loose or unsupportive even on the tightest setting, it may need replacing. If you notice the cups are being stretched or feel too big, the straps keep sliding off and need constant adjusting, the underwire pokes out (possibly even digging into the skin or breast tissue), or the fabric is broken down (for example, you may see small elastic fibers sticking out), your bra may be past its sell-by-date. If you are getting measured once or twice a year, it may also be worth doing a bra inventory every six to nine months to check whether or not your bras are wearing out.

## How do I care for my bras?

Bras can be expensive, so it is worth taking care of them to keep them in good condition. Many bras come with washing instructions to hand wash only; this is because wires and hooks can damage a washing machine or decorative lace and other delicate fabrics. If you use a machine, put your bras in a lingerie bag to avoid them getting tangled in a mixed washing load, which can stretch bras and straps. Avoid using a tumble dryer as the heat can affect the bra; instead, line dry by hanging the bra over its central gore, or lie flat on a drying rack. Hanging bras by the straps can cause the straps to stretch due to the weight of the wet material.

How often you wash your bras is a personal choice, as not all "wears" of a bra are equal—a sedentary day in a cool office environment is not the same as a hot, sweaty hike for a couple of hours on a summer's day. Don't leave it longer than a couple of wears though, as the salt in sweat can break down bra elastic over time.

It is worth having a few bras on rotation, as this can help extend the bra life by allowing the elastic fibers to return to their normal length before being stretched and held under tension during wear. A well-fitting bra is likely to last longer than a poorly fitting one that is stretching to fit you.

Storing your bras properly can also help avoid stretching the fabrics. Ideally, lie bras flat in a drawer, or twist one cup inside another. Avoid hanging bras by the straps.

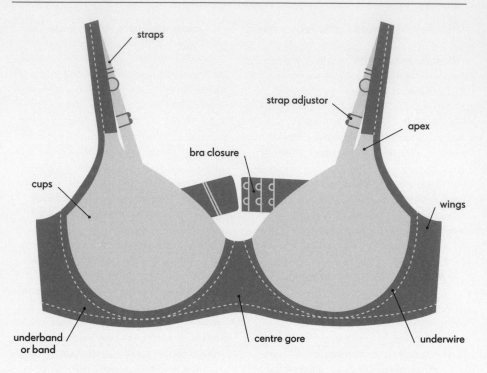

straps

strap adjustor

apex

bra closure

cups

wings

underband
or band

centre gore

underwire

## Bra structure

When you are shopping for a bra, it can be useful to know the parts of the bra and what they do. For a piece of clothing mostly hidden under clothes, bras are a masterpiece in under-appreciated engineering!

### UNDERBAND

• The hero of the bra, the underband carries most of the weight of the breast. It is made up of the wings and center gore and acts like a cantilever, or shelf, to support the breasts.

• The band needs to fit snugly, neither too tight nor too loose, and should be parallel all the way around, not rising up at the back.

### WINGS

• The wings start at the cup and extend round to the back, to join with a closure, such as a hook-and-eye. If the bra is front-fastening, the wings can be one piece.

• The wings should be the correct height for you; if they are too high they can cause rubbing and chafing in the armpit.

(.)(.)

## CENTER GORE

- The part of the band that forms the flat area between the two cups, this should sit flat against the breastbone.

- If the center gore gapes away from the breastbone, the band may be too large, or the cups may be too small.

- Bras will have center gores of varying shapes and heights. Depending on the shape of your breasts you may find that center gores of a particular shape or height are uncomfortable. For example, if your breasts are close together then a bra with a high center gore may be uncomfortable.

## CUPS

- The parts of the bra in which you put your breasts.

- The amount of coverage in the cup will depend on the style of the bra, ranging from full cup to low coverage.

- There may be two separate cups or it may appear that there is only one space for both breasts, as in some sports bras.

- Cups can be molded (for example, made of foam), or padded, or not. Padded cups give more nipple coverage.

## APEX

- The part of the bra where the cups meet the straps.

- The breast tissue extends up to the collar bone, so the apex should fit well and not dig into the breast.

## BRA CLOSURE

- This can be at the front or back. At the back, this tends to be a hook-and-eye closure, and at the front, hook-and-eye, clasps, or a zip.

## STRAPS

- The straps help stabilize the fit of the bra and provide some support.

- These should be adjustable to ensure a good fit on people of different heights.

- The straps should be adjusted so that they are neither too tight and at risk of digging in, nor too lose and prone to slipping.

## UNDERWIRES

- These are not always present in a bra, but if they are, they will generally be in a tunnel or canal in the lower part of the bra cups.

- Wires should fit around the shape of the breast and not cut into the breast tissue.

(.)(.)

# which bra is best?

Fashion, innovations in design, and issues such as cost and sustainability mean that there are more bra styles available to us than ever.

There is no all-purpose bra that suits every context. What you want or need out of a sports bra is different from that of a strapless bra to go underneath a particular outfit, so you may need to buy a few different bras. New developments in technology mean that the choice is greater than ever. If a bra company was—and is still—involved in the production of spacesuits (see page 20), who knows what could be next?

## Seamed vs seamless

Bras with cups made of multiple pieces of fabric are known as seamed bras. They tend to offer more support than seamless bras, so are often useful for larger cup sizes. Balconette and half-cup bras tend to have seams, which shape the breast. Check that the seams don't rub against your skin. Seamless bras are made from a single piece of fabric, which is often stretchy, and tend to be less visible under clothes. It is personal choice, but seamless bras may be more useful for smaller breasts than larger ones as they generally give less support.

## Underwire vs wireless

Underwires (made from metal or molded plastic) in bras give more support and shape to your breasts. The wire should follow the path of the inframammary ligament, but as all humans are different this is difficult to do; if the geometry of the cup of the bra doesn't match your breast it can lead to the wire digging in. Try several on to find the best fit for you (see page 63). Some people prefer the support and shape that underwire bras can create. Others prefer the relaxed comfort of bras without wires. There is no wrong answer.

## Straps vs strapless

Although the majority of the support to the breasts is related to the underband, some support does come from the straps. As the breasts get larger it becomes more difficult to support the breasts fully with a strapless bra unless the underband comes lower down the torso, a bit like a short corset.

## FULL CUP

These offer full coverage of each breast and can feel more supportive if you have larger breasts.

## T-SHIRT

This seamless style has soft cups and is designed not to show under a t-shirt.

## BRALETTE

Soft, wireless, and unsupportive, this style can look like a crop top.

## PLUNGE

A low-cut style that often features padding under the cups to maximize cleavage.

## MULTIWAY

This has straps that can be moved into different positions, like racerback or halterneck.

## SPORTS

Designed to support and minimize breast movement during exercise (see page 77).

( . ) ( . )

## STRAPLESS

Designed to be worn under outfits where you don't want your bra straps to show. The wider band holds the weight of your breasts.

## STICK ON

Without straps or a band, this style of bra sticks to your chest. It offers little support.

## BALCONETTE

This bra has low-cut cups and wider-set straps, giving more uplift and accentuating cleavage.

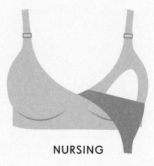

## NURSING

Supportive with drop-down cups to easily expose breasts for breastfeeding.

## POST-MASTECTOMY

Soft seams, a wider band, and separate cups make a post-mastectomy bra a comfortable option post-surgery (see page 183).

( . ) ( . )

# how to measure your bra size

Although bra sizes aren't standardized, having an understanding of how to measure yourself is a useful starting point (then try lots on).

## Finding your band size

Measure underneath your bust, where the band would sit. Breathe out and relax as you do this. The tape shouldn't be too tight or uncomfortable. Make a note of the number.

- Bras sold in the UK and the US have band size in inches. Your bra size is NOT the number you see on the tape. You need to add to the measured numbers: if the tape measure number is even, then add 4, if it is odd, add 5. For example, if you measured 30 inches with your tape measure, your band size would be 34, if you measured 31 inches it would be 36.

- Most bras sold elsewhere in the world use centimeters and may be prefaced by the abbreviations EU or INT, or both EU/INT, which stands for European or International sizing. The band sizes are divided into sections of 5 cm, so 70, 75, 80, and so on. Measure under the bust and then round the number down to the nearest 5. For example, a measurement of 83 gives a size of 80.

- In France and Spain, 15 cm is added onto the European band sizing. In Italy, each 5 cm increment is given a number or a Roman numeral.

- In Australia, although the measurement is taken in centimeters, it is then converted into an AU size. For example, 77-82 cm equates to a size 12, 83-88 cm a size 14, 89-94 cm a size 16, and so on.

| UK/US | 28 | 30 | 32 | 34 | 36 | 38 | 40 | 42 |
|---|---|---|---|---|---|---|---|---|
| EUROPE/INTERNATIONAL | 60 | 65 | 70 | 75 | 80 | 85 | 90 | 95 |
| FRANCE/SPAIN | 75 | 80 | 85 | 90 | 95 | 100 | 105 | 110 |
| ITALY | 00 | 0 | 1/I | 2/II | 3/III | 4/IV | 5/V | 6/VI |
| AUSTRALIA | 6 | 8 | 10 | 12 | 14 | 16 | 18 | 20 |

( . ) ( . )

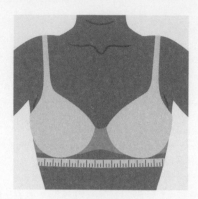

**band size is measured
under the bust**

**bust size is measured across
the fullest part of the breast**

## Finding your bust size

Now, measure around the fullest part of your
breast (generally around the nipple). Don't pull
the tape too tightly—the tape measure should
not dig into your breasts. Breathe out and
relax while you do this, then round up to the
nearest inch or centimeter.

## Calculating your cup size

Now, subtract the band size from your bust
measurement. The difference is the cup size.
But of course, it isn't quite as simple as that.

- In the UK and US, if your band size is 36
  inches and your bust measurement is 40
  inches, the difference between the two
  measurements is 4, meaning that your
  bra size is 36D. (1 = A, 2 = B, 3 = C, 4 = D)

- The UK and US sizing varies for larger sizes,
  with double letters such as DD being more

widely used in the UK. See the chart opposite
to guide you.

- For EU and International sizing you use the
  rounded down underbust measurement to
  calculate the cup size. For example, if your
  underbust measurement is 83 and your bust
  size is 96, the calculation is 96-80, giving a
  difference of 16 (not 13). The cup-sizing systems
  in France and Italy are also calculated this way.

- Bras sold in the EU do not use double letters,
  so the letters increase more rapidly.

- The Australian system is similar to the UK
  system, with double letters being used.

To make things even more complicated, not
all bra manufacturers will follow these sizing rules
because there is no standardization in bra sizes,
Although a one-inch difference between the
underband and bust size is supposed to signify
an A cup, in reality it may not. So, always try
before you buy!

( . )( . )

| DIFFERENCE BETWEEN BAND AND BUST SIZE | | CUP SIZE | | |
|---|---|---|---|---|
| IN INCHES FOR UK/US | IN CM FOR EU/INT | UK | US | EU/INT |
| 0 | 10–12 | AA | AA | AA |
| 1 | 12–14 | A | A | A |
| 2 | 14–16 | B | B | B |
| 3 | 16–18 | C | C | C |
| 4 | 18–20 | D | D | D |
| 5 | 20–22 | DD | DD/E | E |
| 6 | 22–24 | E | DDD/F | F |
| 7 | 24–26 | F | G | G |
| 8 | 26–28 | G | H | H |

## • BRA SIZING VARIATIONS •

If measuring for bras sounds complicated, then that is because it is! Globally, the system is not standardized, and as so many of us now buy online, things are even more complicated. The UK standard measuring system itself appears to be inaccurate and underestimates bra size, generally suggesting a cup size that is too small and a band size that is too big.

Sizing for larger breasts is even less accurate, as the method was devised for sizes up to a D cup, but sizes now go up to an M. Use the measurements as a guide, and then focus on the fit of the bra for you, rather than what the label says.

(.)(.)

## Choosing the right bra for you

Now that you have figured out what you think is your size is, it's time to try on some bras. Here are some tips to help you get the right fit:

• Once you have put on the bra, fasten it, bend forward slightly at the waist, and lift your breasts to put them fully in the cups.

• Ideally, the band should be secured on the outermost, loosest hook (the "largest" position). All bras will loosen as they age, so you will be able to keep the band taut and fitting correctly by moving along the hooks.

• The band should be level all the way around, meaning that the back of the bra should not rise up but be level with the front of the band. This may be lower than you think.

• The bra should fit snugly, neither too tightly, which pinches the skin, nor too loosely. You should be able to slide one finger underneath the band, but no more.

• Adjust the shoulder straps to the right height; not digging in, but not too loose. Again, these might loosen over time, so you will need to adjust your shoulder straps fairly often.

• Now check in the mirror. Your breasts should fit into the cups, without bulging, and the cups shouldn't gape above the breasts.

• If the bra has underwires, they should sit flat against the skin resting against the rib cage, not digging into the breast tissue or armpit.

• The center gore (see page 55) of the bra, should sit flat against your breastbone.

• Breasts aren't symmetrical, so you may need to have one strap shorter than the other, or choose a bra made of a fabric with some stretch in it to compensate for a larger cup size on one side. Some bras have small pockets in the cups for padding and this can be removed on one side or both.

For information about bras for pregnancy, breastfeeding, and post-breast surgery, see page 183.

( . )( . )

# does my bra fit properly?

Ensuring that your bra fits well is essential for comfort. Your breasts change throughout your life, so you will need to check bra fit regularly.

The huge variety of sizing standards can make shopping for bras a frustrating and costly experience. You do need to give yourself time to try on a range of styles and sizes, to make sure you get bras that are right for you.

The good news is that it is easier than ever to check that you are wearing the right-fitting bra. You can now book virtual fittings online, so you don't even need to leave the comfort of your home. Many stores offer free in-store fittings with a trained bra-fitter. You can keep your clothes on if you prefer, and there should be no pressure to buy.

In most countries, bra sizes are written as a number and a letter, for example, 34C or 75D, with the number being the band size and the letter the cup size. But the numbers vary from country to country, and even manufacturer to manufacturer, in what they mean. Even different bras produced by the same brand may fit differently.

The size of your bra does not matter; no one else can see the label, and even if they could, so what? It is the fit of the bra which is important. However, it is useful to have an understanding of bra sizes and how you might measure yourself as a starting point, but after that it is important to start trying them on. Don't worry that the sizing sounds complicated and different between countries—that is because it is—but all that really matters is that your bras fit you.

---

### • SISTER SIZING •

The cup size is not actually the volume of the breast. When people describe themselves as a "D" cup, this actually refers to both the band and cup size, in combination. So a 34B bra is bigger than a 32B bra, even though the cup size is the same. This leads to the concept of "sister sizing"—that is, if you find that your bra is too big on the cup you can move up a band size, and down a cup size. For example, if a 36D bra is too big on the cup, then move to a 38C, decreasing the cup size and increasing the band size to compensate. Conversely, if a 36D is too small on the cup you can move to a 34E, decreasing the band size as you increase the cup.

(.)(.)

## SIGNS OF A POORLY FITTING BRA

### BREAST BULGING OVER TOP

If your breasts are bulging over the top of the cups you may need a larger cup size, or a different style bra that covers more of the breast.

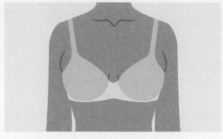

### BREAST BULGING AT SIDES

Breasts bulging out of the sides of the cups may mean the shoulder straps are too short, or you may need a larger cup size.

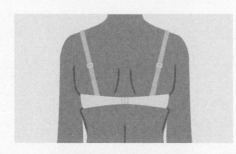

### BRA CAUSING DISCOMFORT

If the underwire, band or gore is pinching and causing discomfort, it may be that the bra isn't sitting in the way it was designed to sit. The band might be too small or too big, so the bra rides up, or the cup is too small for the breast. Changing the band or cup size should help.

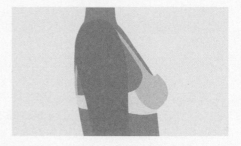

### GAPPING CUPS

This occurs at the apex, where the cup meets the strap. Tightening the shoulder straps may help, or a smaller cup size may be needed. If there is room for something else in the cup, apart from your breast itself, it is too big.

( . ) ( . )

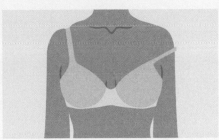

## BAND TOO HIGH AT BACK

The band should be parallel at the front and back. If it is riding high at the back then the shoulder straps may be too short and need loosening. Alternatively, the band size is too large, meaning that the breasts are unsupported and are pulling the bra down at the front, so that it rises up at the back. Or the band size may be too small for you.

## STRAPS TOO TIGHT OR TOO LOOSE

Shoulder straps that are too tight or dig in can be made looser. Over time, straps lose elasticity, becoming looser and prone to slipping off the shoulders. Adjustable straps may help compensate for this. If you slide your thumbs under the shoulder straps there should be a little give, but you shouldn't be able to lift the straps higher than a few centimeters.

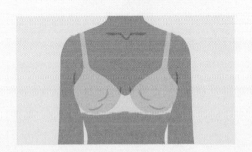

## WRINKLING CUPS

If the fabric and structure of the bra is wrinkling in the cups, a different cup size may be needed, so that the fabric lies smoothly over the breast.

## GORE ISSUES

The center gore should sit flat against the chest. If it digs in, then the band size may be too small. If it gapes away from your chest, the cup size may be too small or too big.

(.)(.)

# buying a bra

## WHAT HAPPENS AT A BRA-FITTING APPOINTMENT?

One of the easiest ways to ensure that your bra fits correctly is to make an appointment with a bra-fitter at a store. Many stores offer this service, free of charge, though you may have to book in advance. You don't have to get naked or show the fitter your breasts. Instead, you will be asked to take off any bulky top layers until you are in a tank top, or if you are comfortable, your bra. The fitter may take measurements, guide you through taking your own measurements, or assess by eye. Then they will help you select bras and assess how they fit, waiting outside until you are changed into each bra.

————

## SHOULD I BUY BRAS ONLINE?

Some sellers and stores will offer virtual bra-fitting appointments to help you choose a bra that fits and gives the support and appearance you need. You are likely to need to try multiple sizes and styles so be sure to check the returns policy and whether there is a fee for mailing returns before you buy.

————

## WHAT ABOUT BUYING SECONDHAND?

Bras can be bought secondhand and many available will be of good quality with plenty of life (and elastic!) left in them, perhaps bought and the person changed size, or bought for a specific outfit. As always, fit matters. If you would like to donate old bras, they can be sent to various charities and organizations.

————

## IS THE PRICE WORTH IT?

When buying a bra, there are multiple things to take into account, from the size, fit, style, and function of the bra, to the sustainability and cost of the item. Another important consideration is the appearance of the breasts when the bra is on, both with and without clothes. How much you want to spend is a personal choice that will depend on all of these factors, and more.
No matter the price, the bra should fit well.

———

## DO I NEED A DIFFERENT BRA DURING PREGNANCY?

Changes in your breasts during pregnancy, such as increased size and expanding rib cage, mean that it is likely you will need a different bra size than before, if you choose to wear a bra. A bra-fitter can help with ensuring that you are wearing the correct bra size. There are bras specifically marketed for pregnancy, though these are not essential; some people prefer to wear a sports bra for extra support. For breastfeeding bras, see page 58.

———

## WHAT ABOUT DURING CANCER TREATMENT?

Your bra size may change during and after breast cancer treatment, so you may need to check your bra fitting. For choosing a bra after a mastectomy, see page 183.

———

# 04

# After puberty

# the menstrual cycle

Changes to how your breasts feel on a day-to-day basis can
be expected, due to cyclical hormone changes taking place.

THE MENSTRUAL CYCLE, DAY-BY-DAY

This diagram is based on an
average 28-day cycle; yours
might be shorter or longer.

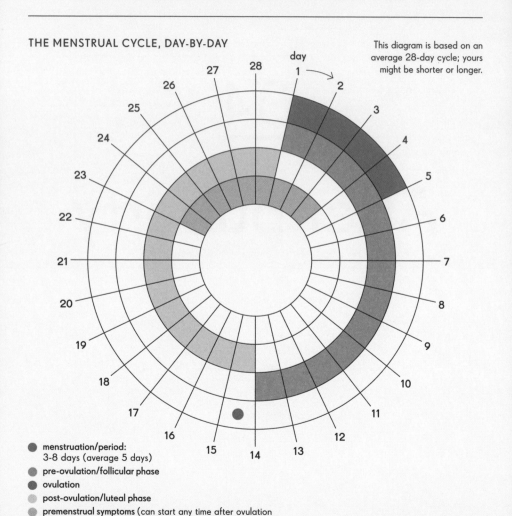

- ● menstruation/period:
  3-8 days (average 5 days)
- ● pre-ovulation/follicular phase
- ● ovulation
- ● post-ovulation/luteal phase
- ● premenstrual symptoms (can start any time after ovulation
  and can last up to about 5 days after menstruation)

(.)(.)

Your breasts might feel very different during the various stages of your menstrual cycle; on some days they might feel tender or sore, or you might need to adjust your bras to feel more comfortable. What's going on?

Breast symptoms leading up to your period affect around half of us and can include breast swelling, tenderness, and pain. You may notice that your breasts also become more lumpy in the lead-up to and during your period, due to the stimulation of the milk glands; this returns to normal after your period. Some people describe the lumpiness as like tiny cobblestones, or like frozen peas or corn under the skin, and it can be extremely tender (for more information on cyclical breast pain see page 128).

## Understanding your cycle

The menstrual cycle starts on the first day of your period, and even as you are menstruating, your brain is gearing up for the next cycle. An area of the brain called the hypothalamus releases gonadotrophin-releasing hormone (GnRH). This stimulates another part of the brain, the pituitary gland, to release follicle-stimulating hormone (FSH). FSH stimulates the ovaries to produce estrogen, and the development of one or more follicles follows, producing an egg (or sometimes two).

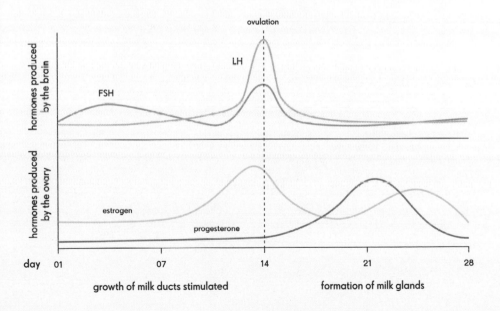

( . ) ( . )

## BEFORE OVULATION

In this first half of the cycle, known as the follicular phase, the levels of both FSH and estrogen continue to rise. Estrogen has multiple roles in the body, including building up the lining of the womb ready for implantation (if fertilization occurs). If your cycle is 28 days long, on about day 12 or 13, the rising level of estrogen stimulates the pituitary gland to release another hormone, luteinizing hormone (LH). This LH surge then triggers ovulation, on day 14 in a 28-day cycle. The length of your menstrual cycle is related to the length of this first, follicular phase, as your next period will start approximately 14 days after ovulation.

## AFTER OVULATION

After ovulation, during the luteal phase, the shell of the follicle which produced the egg, known as the corpus luteum, produces a further hormone, progesterone. Just like estrogen, progesterone acts on multiple parts of the body; one role is to mature and stabilize the lining of the womb, ready for implantation. Unless conception occurs, after about a further week, the levels of both estrogen and progesterone fall, leading to the womb lining becoming unstable and being shed—your period. The whole cycle then starts again.

# What happens to your breasts

The follicular phase is estrogen-dominant, while the luteal phase is progesterone-dominant; these hormones cause changes to your breasts throughout the month. Estrogen stimulates the growth of milk ducts in the breast in the first half of the cycle. In the second half, progesterone is involved, stimulating the formation of milk glands and breast lobules, essentially getting ready if conception occurs. These hormonal changes may lead to breast discomfort, often most evident just before your period starts, and improving during or shortly after your period. Levels of both estrogen and progesterone are lowest at the start of the menstrual cycle, so you may notice that your breasts are softer and smaller at this point.

Premenstrual breast-swelling and symptoms are often linked with premenstrual syndrome (PMS) and fibrocystic breast disease, which are benign (not cancerous) changes (see page 139).

(.)(.)

# contraception

If you are considering using contraception, it can be useful to know how this might influence how your breasts feel.

Your choice of contraception can (but not always) have side-effects for your breasts, including temporarily changing their size and increasing their sensitivity. This is because of the hormones in the medication. Some forms of contraception, such as the condom, copper coil, or diaphragm, do not have any effect on the breasts, as they do not contain hormones.

## Hormonal contraception

The combined oral contraceptive pill ("the pill"), contains both the hormones estrogen and progesterone, while other forms of contraception such as the progesterone-only pill (the "mini pill"), implant, injection, and intrauterine system (hormone coil), contain only progesterone.

On starting hormone-containing contraception you may notice that your breasts feel larger than previously, or look swollen. This is due to the effects of the hormones they contain on the breast tissue itself, and also because hormones can cause fluid retention.

Any increase in breast size related to hormonal contraception is not permanent; the breasts will either return to their normal size after a few months or when you stop taking the contraception. Some people may notice that when they are on their pill-free break or inactive pills (7 days in a packet of each 28 days, though you may have a 4-day break or no break at all in tailored pill-taking) their breasts decrease in size again. The hormones in hormonal contraceptives can also cause breast tenderness or pain. This often settles down after a few months, but if it does not or is causing concern, please speak to your doctor about alternative options. For more information on methods to manage breast pain, see page 128.

(.)(.)

# physical activity

For many, breast size and potential discomfort can be a barrier to exercise. Let's look at the science behind breast movement, and how you can best support your breasts.

We all know that we should exercise; the benefits both on your physical and psychological health are varied and many. But how many of us have danced, jumped around, or had to run for the bus and found ourselves clamping our arms across our chests to stop the bounce? When your body moves, your breasts move, too. Understanding this motion, how it impacts us, and what we can do to support our breasts is the key to exercising comfortably (and not avoiding it altogether).

The impact of breast movement and support during exercise should not be underestimated: people with larger breasts report that they choose to participate in less high-intensity activity than those with smaller breasts, and that they consider their breast size to have an impact on the exercise they do. Breast issues can stop people from taking part in physical activity entirely—the stats vary, but in one study, 17 percent of adult women—almost 1 in 5—reported that their breasts are a barrier to exercise. In fact, breast issues, in another study, were quoted as the fourth-largest barrier to physical activity (the top three reasons being motivation, time barriers, and health). However, the same study showed that women who reported higher levels of breast-health knowledge were more likely to exercise. This is hugely important, because it means that by educating people about their breasts, you empower them, leaving them more likely to participate in physical activity.

## • BREAST PAIN AFTER EXERCISE •

More than 70 percent of exercising women report breast pain after exercise. While this is thought to be related to the movement of the breasts, the exact cause is not yet known. The breasts can move a lot, in various directions (see right), and these repetitive movements can lead to changes to internal breast structure.

## How your breasts move

Breasts move during physical activity, and not just in an up-and-down motion. In fact, they move in all directions. Your breasts also move independently of one another, rather like a sideways figure-eight or butterfly motion. As you walk, the movement of the torso causes the breasts to move up-and-down, forward and backward, and side-to-side. Even when just walking breasts can move up to about 4 cm.

## BREAST MOVEMENT AND SUPPORT DURING EXERCISE

Here is an example of slow, gentle jogging, while wearing no bra, a low support bra, and a high support bra.

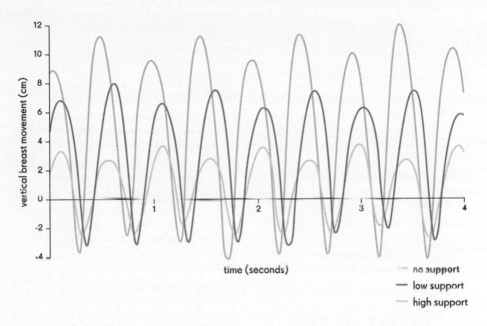

time (seconds)

····· no support
— low support
— high support

# What happens during exercise

Studies have shown that breasts can move about 12-15 cm during high-impact exercise such as running, and this can increase further in activities such as jumping jacks, up to almost 20 cm. Breasts can be heavy, too, further increasing the effect of movement.

The breast is made up of soft tissue, fat, and glandular tissue and has limited internal support. The breast is supported by connective tissue called Cooper's ligaments and the skin (see page 26). The repetitive movement of exercise is thought to stretch the Cooper's ligaments over time, which can contribute to

breast sagging (breast ptosis), though currently there is no research to prove this. Movements during exercise also affect the skin, meaning that there is less support. As you get older, in particular after menopause, there are changes to the skin's elasticity which may result in sagging, but exercise may contribute to this happening earlier in life.

Breast changes and pain related to exercise can occur in all breasts shapes and sizes, but it seems clear that there is more movement in larger, heavier breasts. The type of exercise done is also key, with high-impact exercise such as running resulting in larger movements than low-impact exercise such as Pilates.

## Adolescence

The number of females reporting that their breasts are a barrier to exercise is even higher in adolescents. A study of over 2,000 girls aged 11-18 showed that 46 percent avoided physical activity due to their breasts. This figure is 51 percent among 13–14-year-olds, and higher still for those with larger breasts (63 percent). This may not simply be due to breast pain, but also factors such as embarrassment related to their breasts and breast bouncing. Add this to the fact that periods are reported as the most common barrier to participation in sports, and we have the statistic that around 1 in 3 girls drop out of sports by puberty (on average, around age 12). More than 60 percent of females don't do any exercise at all.

About 87 percent of participants in the study said they wanted to know more about breast health. While there are other factors involved, breast education and support could improve these low participation numbers, which could then have a positive impact on long-term health.

## We need more research

There is so much more work and research to be done on breast health during movement. In comparison to the breadth of studies into sports shoes, there is little on breasts and sports bras. Of the research done on breast movement in athletes, much of this relates to the effects of running, but the movement of the breasts will be different in each sport.

We also need research into the impact of surgery, such as mastectomy or breast-conserving surgery; breast biomechanics (movement); the impact of menopause on breasts; and far more, in order to be able to inform sports-bra development.

---

### • PERFORM AT YOUR BEST •

Breast movement during high-intensity activities can lead to biomechanical and physiological changes, which can then have an impact on performance. For example, if there are changes in muscle activity in the upper body, stride length may become shorter, and changes to breathing patterns are seen.

How does this happen? Put simply, if your foot hurts when running you might adjust how you run, or limp; if your breasts hurt you will also make changes to your movements that you may not even be aware of, to try to reduce discomfort. If you want to perform at your best, whether an amateur or professional athlete, well-fitting bras are essential.

---

(.)(.)

# sports bras

Good breast support during exercise reduces breast pain and may help keep you active, so let's look at choosing a sports bra.

## Do I need a sports bra?

It is up to you, but they are helpful. Wearing a sports bra can reduce breast movement, leading to less breast pain, whatever age and breast size you might be. It may even improve your performance (see page 76).

## Which sports bra is right for me?

There are lots of factors to consider when choosing a sports bra, including the amount of support you need, fit and comfort, as well as how the bra looks, and how much it costs.

Sports bras can be classified by impact of physical activity: high-, medium- or low-impact, with the implication that a high-impact bra is suitable for a high-impact activity such as running, while a low-impact sports bra would be more suitable for an activity such as yoga or Pilates. However, there is currently no regulation around this classification, meaning that different companies can label their bras as high-, medium- or low-impact support without the bras being independently tested against other models. Some companies will carry out internal or external testing, but they don't have to.

It is also important to check that the bra fits (see page 79) and is comfortable. After all, there is little value in buying a bra so supportive it is actually constrictive and uncomfortable to wear;

you are less likely to wear it. Conversely, buying one which is extremely comfortable but doesn't provide the support you need isn't helpful either. This balance will be different for everyone and even for the same person, depending on the activity you are doing. In general terms, you will need more support if you have larger breasts, and as the impact level of the sport increases. Research suggests that encapsulation sports bras (see page 78) may work better for people with larger breasts (over a D cup) to reduce vertical movement. Compression bras seem to be more suitable for smaller breasts. A combination bra which both elevates and supports the breasts, and also compresses, is generally suitable for most wearers.

---

### • THE FIRST SPORTS BRA •

The first sports bra was created by Lisa Lindahl and costume designer Polly Smith in 1977. It was initially called the "Jockbra," because it was made out of two jock straps sewn together. The name was later changed to "jogbra," inspired by Lindahl and her sister running.

---

**ENCAPSULATION BRA**

This design supports each breast individually within cups. The style has additional features such as padding for support and wide shoulder straps. These bras are generally sized similarly to everyday bras, with a band size and cup size (see page 59).

**COMBINATION BRA**

This combines elements of both encapsulation and compression styles.

## Pregnancy and breastfeeding

If you want to exercise while pregnant, a well-fitting sports bra can help support your changing breasts. While the function of the sports bra (to support the breast during physical activity) remains the same during pregnancy, you might need a larger size, or one that offers less compression so your changing breasts feel comfortable. When breastfeeding, the sports bra may need to have cup-release straps so that the breast can be accessed for feeding.

Pregnant women, and those who are breastfeeding or who have been pregnant in the last year, are generally excluded from research in sports bras. This is probably for ethical reasons, as the research is likely to involve the woman performing activities in a laboratory setting, such as running on a treadmill, with and without a bra on, or wearing different bras. However, it is important that research is carried out as we know that a lack of breast support is a reason why women stop exercising in general.

**COMPRESSION BRA**

These bras support the breasts by compressing them against the chest. These tend to be sized more generally as small, medium, and large. Compression sports bras may be better for smaller breast sizes, as they seem to be less effective at the larger sizes. They are sometimes also known as crop tops.

In the meantime, while we wait for better research, continue exercising, and wear a well-fitting, supportive sports bra while doing so. Wearing a bra which is not too tight nor too unsupportive should not impact on either the pregnancy or breast development. Moderate exercise does not affect milk supply. If you are concerned about leaking during exercise, you could consider wearing nipple pads.

## Is my sports bra a good fit?

You can use a professional bra-fitter (see pages 52-53) for sports bras, too, if you like. This is perhaps even more important if you do a lot of training. If buying on your own, check the underband, cups, straps, neckline and panel fit. Try doing some jumping jacks and stretches in the changing room. Think of it like when you walk around in new shoes—why not test out your sports bra, too?

## Anatomy of a sports bra

### UNDERBAND

• The main support system of the bra comes from the underband, meaning that the breasts are mainly supported from below, as opposed to being held up by the straps from above.

• Choose an adjustable underband so you can find the balance between support and comfort. Ideally there should also be some flexibility in the underband so that it isn't too constricting while giving breast support.

• The choice of how the underband is secured, such as hooks-and-eyes or zippers, comes down to personal preference, but ensure that the fastenings are covered with material so that there is no chafing or risk of damage to the skin.

• You may prefer to have a bra which opens at the front or the back; for example, a front-opening bra may be easier if you have shoulder-mobility issues.

ADJUSTABLE STRAPS TO
ADJUST FOR HEIGHT AND
FOR STRETCH OVER TIME

UNDERWIRES (IF PRESENT),
TO FIT UNDER THE BREAST
AND NOT DIG IN

ADJUSTABLE FASTENING
TO ADJUST FOR STRETCH
OVER TIME

( . ) ( . )

- While I recommend that you check your bra fits well with a bra-fitter, you can check yourself that the underband fits comfortably and securely by raising your arms above your head when you try it on; the band should not rise upward or gape. Neither should the band pinch the skin.

## SHOULDER STRAPS

- Although the majority of breast support comes from the underband, some support is given via the shoulder straps.

- Adjustable shoulder straps are a must, so you can adjust them to fit you properly. After all, two people can have the same breast size, say 34C, but be very different in height, so won't have the same distance between the shoulder and the breast.

- You can buy classic or vertical, crossover, or racerback strap styles. If the straps and bra fit well, the orientation of the straps does not significantly affect the support of the breast and so can be up to personal preference, or perhaps you find one style easier to get on and off. A racerback or crossover style may help prevent the shoulder straps slipping.

- When you adjust the straps, take care to adjust each strap for each breast, as opposed to doing one and then simply adjusting the other to match. Your body and your breasts are not symmetrical, so adjust the straps accordingly.

- Women with larger breasts often report that a wide vertical strap is the most comfortable style, less likely to cut into the shoulders.

- A nonadjustable high-impact sports bra is likely to stretch over time, and the straps cannot be tightened to compensate, so adjustable-strap bras are likely to last longer.

**A RACERBACK OR CROSSOVER STYLE MAY PREVENT SHOULDER STRAP SLIPPAGE**

**A HIGHER NECKLINE GIVES MORE SUPPORT**

**MOISTURE-WICKING FABRIC TO STAY COMFY AND DRY**

## CUPS

- The cups in a sports bra (or front panel if not molded) should completely encase each breast, unlike other non-sports-bra styles such as a balconette.

- There should be no bulging of the breast tissue itself as this would imply that the cup is too small. Conversely, if the cups are too big there may be gaping or wrinkling.

- The central gore should sit flat against the breastbone and not gape, which would imply that the underband is too big or the cup size is too small.

- Cups can have underwires, seams, padding, and panels. Underwires should sit flat against the ribs and not on the breast tissue, so they fit the shape of your breasts and don't dig in.

- Padding can be supportive but also may function as breast protection during sports which have an element of contact, such as basketball, soccer, and tennis.

## NECKLINE

- The shape of the neckline is not only for aesthetics, but also has a role in the function of a sports bra. This is because breast tissue extends up to the collarbone (see page 26). Generally, the higher the neckline, the more supportive a bra will be.

- Higher necklines tend to reduce the up-and-down movements of the breasts. For every 1cm increase in neckline height, there is a decrease in breast bounce and movement by 0.75 percent.

## SIDE PANELS

- The side panel of the bra is the part which sits under your armpit. Breast tissue extends up to the armpit, so the side panel needs to be wide enough to cover all of the breast tissue.

- Side panels should fit snugly, but not too tightly or digging in, and must not be too high as this can cause rubbing and chafing as you move your arms.

---

### Q: HOW OFTEN SHOULD I BUY A NEW SPORTS BRA?

( . ) ( . )

**A:** *Sports bras may need replacing every six to twelve months, earlier if you notice they are not as supportive as when new. You may also need to buy a new bra if you have gained or lost weight. The frequent washing of sports bras affects their elasticity and support; evidence shows that even after as few as 25 washes there is a reduction in support. This is worsened by the effects of actually wearing and exercising in the bras.*

( . ) ( . )

# chest binding

Chest binding is a practice sometimes used to flatten and reduce the size and appearance of the breasts.

Chest binding, also known as breast binding, involves using some form of compression to bind the breasts close to the body so they look and feel flatter. For some transgender and nonbinary people, the breasts are a source of gender dysphoria (the emotional distress you can experience if your body doesn't match your gender identity), and chest binding can help alleviate these feelings.

Binding has been shown to have positive effects on the mental health of people experiencing gender dysphoria, such as increasing self-esteem and decreasing anxiety. In one study, 70 percent of participants experienced a positive mood after starting binding, compared to 7 percent before.

There are, however, health risks linked to binding. Research has shown that approximately 90 percent of people experienced at least one negative side-effect from binding. These include: skin irritation and tenderness, broken skin, skin infections, bruising, scarring, itching, back pain, and damage to the ribs, as well as difficulties breathing and issues with overheating, especially in hotter weather. Wearing a binder for longer periods of time is associated with an increased risk of side effects. If you have an underlying health condition such as scoliosis or a respiratory condition such as asthma, please seek medical advice before considering binding.

## Types of binders

You can buy binders online and in stores. Various options are available but have similar risks and benefits.

### CHEST/BREAST BINDERS

These are pieces of clothing designed to flatten the appearance of the chest area and come in various sizes and styles. They are available in short and long lengths, to cover the breasts or the breasts and torso.

### SPORTS BRAS

While sports bras do not provide as much compression as a binder, compression-style (see page 78) sports bras can also be used.

### ELASTIC THERAPEUTIC TAPE

Also called kinetic or kinesiology tape, this is a form of medical tape that is often used in physiotherapy, and can be used to bind breasts.

(.)(.)

## Things to remember

• Ensure that the binder fits well, meaning that it is not so restrictive that you cannot breathe properly. It should be able to stretch slightly to allow you to take deep, full breaths.

• A chest binder should not be worn for longer than eight hours consecutively.

• Do not wear more than one binder or type of binder at a time.

• Do not use other kinds of tape apart from kinesiology or kinetic tape. Sticky tape (such as Sellotape and Scotch tape), duct tape, and electrical tape can irritate the skin and restrict the movement of the chest and breathing.

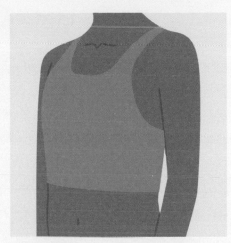

**HALF BINDER**

• Avoid using a binder during exercise and physical activity because you need to be able to move easily and breathe deeply.

• Do not wear a binder to sleep in.

• You may be advised to avoid wearing a binder, or to use a binder for shorter periods of time, if you are planning to have top surgery to remove breast tissue. This is because prolonged use of a binder can have an impact on the skin.

**FULL BINDER**

• If you are uncomfortable, or have pain or difficulty breathing, then remove the binder. If the symptoms don't resolve on removing the binder, please seek medical advice.

# nipple piercing

Are you considering a nipple piercing (when a nipple is pierced and jewelery inserted)? Let's look at safety and any potential issues.

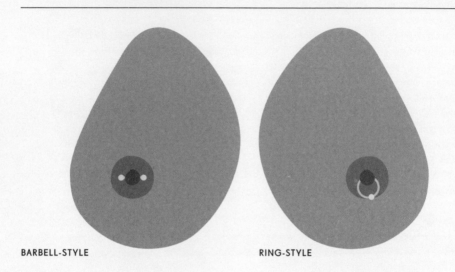

**BARBELL-STYLE**                                    **RING-STYLE**

Piercing has been practiced for thousands of years, from Ancient Rome to North America's Indigenous peoples, such as the Karankawa people. In Victorian times, some considered it fashionable to have pierced nipples, perhaps linked to when lower-cut dresses were in vogue.

There are plenty of reasons why you might want a nipple piercing; you may simply like how they look and feel, but piercing can also be used as a treatment for inverted nipples (see page 38). Some people report that their nipples become more sensitive after piercing, increasing sexual pleasure, while others report decreased sensitivity after piercing.

Nipples are packed with nerve endings, so the process of piercing may be more painful than other areas. The pain can be short-lived or last a few days afterward. Although the nipple piercing may look healed on the outside, they actually take 9-12 months to fully heal, so it is important to continue with your piercing aftercare for this time. Changing the jewelery too early can damage the healing inside of the piercing.

You are advised not to get a new nipple piercing if you are planning on getting pregnant or breastfeeding in the following 12 months. For information about breastfeeding with a nipple piercing, see page 102.

(.)(.)

## Things to remember

**01. Choose your provider carefully.** Always select a provider who uses sterilized equipment. It is possible for the nipples to reject the jewelery (see right), but this is less likely if titanium or gold is used. Piercing professionals tend to recommend that the nipple is pierced and a barbell-style piece of jewelery inserted, as this allows the nipple to swell.

**02. Clean and protect your piercing to aid healing.** Clean the piercing twice a day using a saline (saltwater) solution. You might choose to wear a well-fitting bra or top to ensure that the jewelery doesn't snag on anything and cause trauma to the skin. Avoid touching and stimulating the nipple while it is healing. Smoking can delay skin healing. You can use over-the-counter painkillers and cold compresses to help with any pain. There is likely to be some swelling after the piercing, but if this is severe or persists, then speak to your doctor.

**03. Check for discharge and infection.** In the first few days after the piercing there may be some clear or white/creamy-colored fluid oozing from the site, which may be blood-stained. If this becomes green, offensive in smell, or if bleeding becomes worse or persists, speak to your doctor. The fluid may form a crust on the nipple, which can be itchy—don't pick it. Instead, soak in saline solution until it softens and can be wiped away. If the nipple is infected then your doctor may recommend antibiotics

and potentially leaving the piercing in, as the hole acts like a drain for any discharge, hopefully preventing an abscess forming.

## Bumps and scars

It isn't always possible to entirely prevent scarring and bumps. However, aiming to prevent infection, or treating any infection promptly, can help, as will avoiding any products which might cause irritation, including low-quality jewelery.

Bumps around the piercing can occur, which may be due to hypergranulation tissue, which is an overgrowth of tissue. Irritation, perhaps if the barbell isn't long enough and is irritating the skin, can also lead to bumps. It is also possible to get hypertrophic scarring, where there is more scar tissue than normal, or keloid scarring, where the scars can grow around the pierced area.

## Piercing rejection

Piercing rejection occurs where the body rejects and pushes out the piercing. This is more common with surface piercing than with a piercing which goes all the way through, such as an ear or nipple piercing, but it can occur. Signs include the jewelery protruding more than it should; the piercing becoming red, sore, and irritated; the hole getting bigger; or the jewelery moving more than previously. See your piercing provider for advice as they may recommend removing the piercing.

# 05

# Pregnancy and breastfeeding

# changes during pregnancy

Your breasts will change during each trimester of pregnancy, including in the "fourth" trimester—postpartum—due to fluctuating hormone levels. Here is what you might experience ...

Fluctuations in hormone levels mean that breast changes can occur as early as one week after conception—before you even know that you are pregnant. These hormone fluctuations continue throughout pregnancy and into the postpartum phase after delivery (whether or not you choose to breastfeed). Your blood volume increases by around 50 percent during pregnancy, to supply the fetus with oxygen and nutrients, and this will be evident in your breasts, as well as the rest of your body.

Some of the most common changes to breasts include tenderness and increased sensitivity, along with growth and perhaps stretch marks. Different people will have different symptoms, and you may notice some of these changes more than others, and even between pregnancies.

Please be reassured that noticing, or not noticing, changes in your breasts does not signify any problems or issues with the pregnancy or your ability to breastfeed.

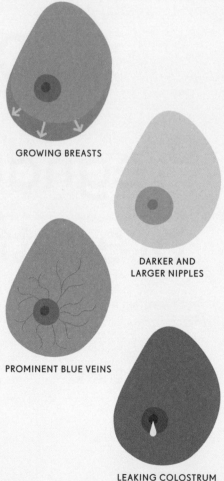

GROWING BREASTS

DARKER AND
LARGER NIPPLES

PROMINENT BLUE VEINS

LEAKING COLOSTRUM

# First trimester (weeks 0–13)

- The levels of estrogen and progesterone rise during pregnancy; there is increased blood flow to your breasts; and the milk ducts and lobules (see page 26) start to develop.

- Some people will notice their breasts grow in the first trimester, while others won't notice any change until later in their pregnancy. Both developments are normal.

- Breast tenderness is often one of the first signs that you are pregnant. You may notice tenderness even before your period is due or before you take a pregnancy test.

- Breasts may feel tender, sore, or heavy. There may be tingling sensations, which may be due to the increased blood flow and the new developments taking place.

- You may only be aware of discomfort or sensitivity if you lie on your front to sleep, or you may not experience it at all.

- Tenderness and discomfort can extend up into the armpit, because breast tissue (the Tail of Spence, see page 26) extends up that far.

- Nipples in particular can feel more sensitive than usual, or touch can feel painful.

- Breast tenderness and discomfort generally improve after a few weeks as your body adapts, though it can recur.

- Your breasts may start to grow even from early on in the pregnancy and may continue throughout your pregnancy (and beyond if you breastfeed). You may notice that you increase in both band and cup size. This may be most noticeable during your first pregnancy, when this happens for the first time.

- As your breasts grow, especially if they grow quickly, the skin has to stretch and you may develop stretch marks. The skin over the breasts can feel itchy and uncomfortable as it stretches; keeping the skin well-moisturized can ease any itching.

- Stretch marks on pale skin tones can be pink, red, or purple at first, but over time will change and fade to become lighter and perhaps silvery in color. With dark skin tones, stretch marks can look darker or lighter than the rest of the breast skin.

- The amount of blood in your body is increasing, and blood vessels will dilate (get wider). This means that you may become aware of more prominent, visible, blue veins on your skin, often on the breasts.

## Second trimester (weeks 14–27)

- The breasts may continue to increase in size with the areolas often growing larger and darker. The color often returns to the prepregnancy color after delivery and/or breastfeeding, but they may also remain slightly darker than before pregnancy.

- Montgomery's tubercles (the small bumps on the areola, see pages 32-33) may become more prominent during pregnancy. These are the glands which produce antibacterial lubrication for the skin and nipples, ready for feeding. The smell of the secretions produced by these tubercles may even help babies find the nipple to latch on.

- You may notice some nipple discharge of colostrum (the first milk your breasts produce, see page 96), but this is more common in the third trimester.

### • BREAST LUMPS •

You may notice that your breasts become more lumpy during pregnancy. It is important that you check any lumps or bumps with your doctor. Benign causes of breast lumps in pregnancy include fibroadenomas (solid breast lumps, see page 139), blocked milk ducts (see page 108), and galactoceles which are cysts filled with breast milk (see page 111).

## Third trimester (weeks 28 to birth)

- The breasts continue to grow, becoming heavier. The nipples and areolas can continue to grow in size and become darker in color.

- The breasts are able to produce milk; estrogen stimulates the development of the milk ducts, and progesterone stimulates the development of the lobules or milk glands (see page 95). Estrogen also has a role in stimulating the production of the hormone prolactin, which further stimulates breast growth and the production of milk.

- Colostrum (see page 96) may be produced and may leak from the nipples. Colostrum is thick and yellowish in color and is high in nutrients and antibodies and has a role in the development of the baby's immune system. Colostrum or nipple discharge during pregnancy is more common after nipple or breast stimulation but can occur on its own.

- Stretch marks can continue to develop or become more noticeable. You can use a simple moisturizing product, if you feel the skin on your breasts is dry or itchy.

( . )( . )

# Postpartum (after delivery)

- After delivery of the baby, the levels of estrogen and progesterone fall relatively rapidly, while the levels of the hormone prolactin continue to rise. Whether you choose to breastfeed or not, your breasts will undergo changes during this stage.

- The breasts produce colostrum after delivery. Between five days and two weeks after delivery, this changes to milk. The breasts may further increase in size, or stay at their larger size due to the production of milk.

- If breastfeeding doesn't occur, then the breasts often return to their prepregnancy size, though this is also dependent on weight gain. The color and size of the nipples and areolas may (or may not) also return to their prepregnancy state, though this can take months.

- If you breastfeed, then the breasts may return to their original shape and size once you stop, but they may also remain larger than before—either is normal.

- Some people report that their breasts sag more after pregnancy and feeding. This may be more likely if the breasts were larger in size before pregnancy; if you are overweight or obese; or be related to significant weight changes during pregnancy. This is more likely if you have had multiple pregnancies. It is also more likely if you smoke. Moisturize regularly to help the skin's elasticity.

---

### • HOW TO EASE PREGNANCY BREAST SYMPTOMS •

- A well-fitting, supportive bra can help with breast discomfort (see page 92).

- Disposable or washable breast pads can be useful if there is leakage of colostrum.

- Dry, itchy skin on the breast may be soothed with emollients such as emulsifying ointment.

(.)(.)

# maternity and nursing bras

It is likely that your breasts will change size and shape during pregnancy
and feeding, so comfortable bras will be a worthwhile investment.

Breast changes during pregnancy and beyond
mean that it is likely your existing bras no longer
feel comfortable. So, what might you need?

## Maternity bras

There are bras specifically marketed for
pregnancy, although you can shop from other bra
ranges, too; the key is a good and comfortable fit
(see chapter 3). Some people prefer to wear a
sports bra for extra support instead. Wearing
a soft bra at night may help with comfort, too.

A bra to wear during pregnancy should have
good support as well as adjustable straps and
closures so you can adjust as your breasts
continue to change. Wide straps may be more
comfortable (see page 55). If your breasts, or in
particular your nipples, are very sensitive then
choosing bras without seams can avoid rubbing
or chafing. People are often told that they need
to avoid bras with underwires during pregnancy
to avoid blocked milk ducts; but you can wear
underwired bras as long as they fit well (see
page 55) and the wires are not digging into the
breast. You may find that you feel warmer than
usual or sweat more during pregnancy, so
wearing bras made from natural fibers such as
silk or cotton can be helpful. Some maternity
bras double up as nursing bras, with front-clip
openings. These are most useful in the last
weeks of pregnancy but can be used any time.

## Nursing bras

It is likely that your breasts will continue to
change size and shape after birth, so if you are
buying new bras for breastfeeding, try to wait
until you are in the last few weeks of pregnancy.
You may need to switch again afterward if your
breasts change.

A breastfeeding or nursing bra has openings
which allow easy access to each breast, while
still providing some support. Check that the cup
opening clasp is easy to use with one hand as
you are likely to be holding baby with the other.
Wide shoulder straps help with support and, as
always, the bra should fit well, with some
flexibility and space to cope with fluctuations in
breast size.

Your breasts may change size quite quickly
during breastfeeding, so ensuring a correct bra
fit at all times can be difficult. Most nursing bras
don't have underwires, to avoid a wire digging in
if your breasts grow. But there is no evidence
that underwired bras impair milk production
or cause any issues like blocked milk ducts. So
long as your bra and underwire fit correctly, wear
what's comfortable for you.

Wearing a bra at night can help give support
and can be used with nipple pads as there may
be leakage of milk at night.

# breast, bottle, or both?

Whether or not to breastfeed is a personal choice, and the reasons behind that choice will be unique for everyone.

The World Health Organization recommends that babies are exclusively breastfed for the first six months of life, but it is important to recognize that some people may find this difficult or it may not be suitable.

## Fed is best

The saying "breast is best" is, in my opinion, both outdated and unhelpful. There is a huge amount of societal pressure and expectation put onto parents that they will breastfeed their children and continue to do so even if they find it challenging. Yes, there are benefits to breastfeeding, but there also needs to be acceptance that it is not always easy or practical to do. The pressure to breastfeed needs to change to support available if you choose to go this route. A stressed, anxious parent, who may even be unwell (in part due to the stresses of feeding) will also have an impact on the baby.

A healthy parent is essential for child health and development. Thankfully, many of us live in a place and time where we have access to clean water and baby formula which can be prepared safely. Instead of the phrase "breast is best," it could be more useful to say "fed is best"—no matter how you choose to do it.

## The benefits of breast milk

There are a myriad of health benefits to both the baby and you with breastfeeding.

### FOR BABIES

Research shows that babies who are breastfed have stronger immune systems, and have the benefits of passive immunity (receiving antibodies through breast milk). This passive immunity helps protect the baby from any infections you have had, meaning fewer coughs, colds, and ear infections. Breastfed babies also have fewer gastrointestinal symptoms, fewer cases of meningitis, lower rates of Sudden Infant Death Syndrome (SIDS), and more.

### FOR CHILDREN

The benefits of breastfeeding continue as the baby gets older; children who were breastfed have lower rates of allergies, asthma, eczema, diabetes, and some gastrointestinal conditions such as Crohn's disease. They also have fewer respiratory infections and dental cavities.

( . ) ( . )

## FOR THE BREASTFEEDER

There are also potential health benefits for you, such as less bleeding postnatally (and therefore less risk of complications such as anemia) and lower risks of breast and ovarian cancers, cardiovascular disease, and more. For some, breastfeeding can also help mental health; the hormones of breastfeeding can help you feel calm and relaxed, as well as promoting bonding with the child.

Others find that they have negative emotions during the let-down reflex (the moment milk is released from the breast, see page 97). For some, the ease of not having to count scoops of formula, and sterilize and prepare bottles is a bonus. It is important to remember that although a natural process, breastfeeding is not always an easy one. Both you and baby have to learn how to feed, and for some, difficulties around feeding or expressing cause real stress and may worsen or contribute to mental-health issues such as postnatal depression.

### • BREASTFEEDING AND CHESTFEEDING •

The terms breastfeeding and chestfeeding both mean to feed an infant with human milk. On the whole, breastfeeding is used in relation to cis women, while chestfeeding (also known as beesting) may be the preferred term of trans or nonbinary parents.

(.)(.)

# breastfeeding basics

In this section we look at how your body produces breast milk, how to feed your baby, and any questions you might have about breastfeeding.

Let's start by looking at how your body makes breast milk, and what it is. The process of milk production starts early on during pregnancy and is called lactogenesis.

Estrogen and progesterone hormones stimulate the development of the milk glands (lobules) and milk ducts in the breasts (see page 27). Levels of the hormone prolactin rise during pregnancy, which stimulates milk production. Very high levels of progesterone in the body mean that although your breasts might produce colostrum before baby is born (see page 96), it is generally only a small amount.

Progesterone levels drop rapidly after the placenta is delivered during labor. This essentially "unmasks" the higher levels of prolactin, which lead to the production of breast milk around 30-40 hours after delivery. It can take a little longer (2-3 days) for you to be aware that your milk has "come in" and for your breasts to start to feel full. Up until this point, milk production is solely hormone-driven, but from now on, although hormones are involved, there is change in that the supply of milk produced is related to how much milk is used. Your breasts will then produce milk for as long as you continue to feed.

## HORMONE LEVELS DURING BREASTFEEDING

— progesterone    — prolactin    — oxytocin

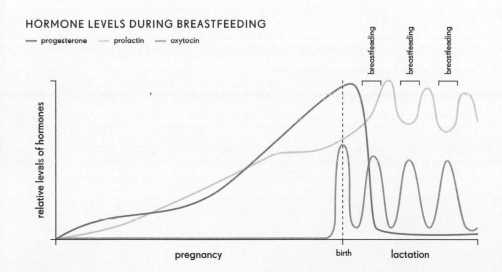

(.)(.)

# Breast milk production

Breast milk changes from colostrum to transitional and mature milk, adapting to suit your baby's needs.

## 01. COLOSTRUM

• Colostrum is produced from about the mid-point in pregnancy. Some women may notice this earlier on in their pregnancy, others in the last few weeks, and others not until the baby is delivered.

• Colostrum is sometimes called "liquid gold" for its nutritional value and color. Rich in protein, vitamins, and minerals, colostrum is thick and golden yellow or cream in color, thanks to high levels of beta-carotene.

• Colostrum is produced for the first few days after birth in small amounts (at this point, your baby's stomach is tiny, about the size of a walnut), generally for 2-7 days. Regular suckling encourages milk production.

• Colostrum is high in immunoglobulins, which are antibodies. This is how your baby gets passive immunity (see page 93), as the antibodies from you are passed to the baby to protect from various bacterial and viral illnesses. For example, adults are offered a vaccine against whooping cough (pertussis) in pregnancy; this isn't to protect them, but to ensure that they have protective antibodies to pass on to the baby.

## 02. TRANSITIONAL MILK

• Transitional milk starts being produced from 2-5 days after delivery, and continues for approximately two weeks.

• Transitional milk is higher in calories than colostrum, to provide for the baby's energy needs. It has high levels of fat and water-soluble vitamins.

## 03. MATURE MILK

• Mature milk is produced from about two weeks after birth for as long as you decide to continue to feed.

• Around 90 percent of mature milk is water, which is needed to keep the baby well-hydrated. The rest comprises carbohydrates, proteins, and fats.

• Mature milk can be white, slightly bluish-white, light yellow, or cream color.

• Foremilk is the milk you produce at the beginning of a feed which is high in water, proteins, and vitamins.

• Hindmilk is the milk that comes after the initial foremilk is released, and has higher levels of fat.

## During breastfeeding

The breasts work on a supply-and-demand basis (see box, below) prompted by suckling. When baby suckles, this stimulates the nerves in the nipple and areola, causing the production of the hormone oxytocin in the brain. Oxytocin triggers the let-down (or milk-ejection) reflex and stimulates the muscles around the glandular tissue in the breasts to contract. These contractions squeeze the milk from the lobules into the milk ducts, and then out through the multiple openings in the nipple.

The let-down reflex can be felt as a tingling, prickling, or even painful sensation in the breast, or a rush of warmth or heat through the breast as the milk comes down. Milk can leak from the other breast, as the let-down reflex occurs in both breasts simultaneously. You may not be aware of the let-down reflex; everyone is different. However, you are likely to notice a change in the baby's pattern of sucking; initially to trigger the let-down reflex the baby sucks rapidly, but as the let-down occurs there is a flow of milk so the sucking changes, getting to a deeper, slower rhythm with swallowing.

Oxytocin is also the hormone released when cuddling that makes you feel good, relaxed, and decreases stress. It seems to help increase the bond with the baby during feeding. Of course, if you choose to bottle-feed your baby, you will also bond with them, often helped by the close contact and eye-to-eye connection.

In the first few days after delivery you may also feel contractions or tightenings in your pelvis, as the hormones linked with breastfeeding also stimulate contractions of the womb to help it return to its prepregnancy size.

You may also feel thirstier than usual, because your body needs extra fluids to create breast milk.

---

### • SUPPLY AND DEMAND •

The breasts work on a supply-and-demand basis, or rather a supply = needs basis. The more that milk is used for feeding (or pumping), the more milk is produced. Conversely, the less milk taken for feeding, the less milk is produced. This is why you can feed multiple babies if you need and choose to.

( . )( . )

# breast milk

### DOES BREAST SIZE AFFECT HOW MUCH MILK YOU MAKE?

No, the production of milk is not dependent on breast size. Different breasts do, however, store different amounts of milk, also unrelated to size. If you have large storage capacity you can hold more milk, meaning you can deliver more milk at a feed, and may be able to go longer between feeds. If you have a smaller storage capacity, you may need to feed more frequently. You may notice that your breasts have different storage capacities, but storage is not the same as production. Whether or not you have a small or large storage capacity you can still produce the same volume of milk, but you may need to feed more frequently.

———

### WHAT IS BREAST MILK MADE OF?

Breast milk contains protein, fat, carbohydrates, vitamins, minerals, and more. The composition will vary between people and even within the same person as it changes to meet the baby's needs at different times of the day. For example, at nighttime, your breast milk may contain higher levels of tryptophan, an amino acid which is converted into the hormones serotonin and melatonin, which make you and your baby feel relaxed and sleepy.

———

### DOES WHAT YOU EAT AFFECT THE TASTE OF BREAST MILK?

Yes! Breast milk is naturally sweet and creamy from the lactose and fat it contains. The food you eat can affect the flavor of your breast milk and may even influence your baby's palate and taste preferences as they grow up. Medications, smoking, and alcohol can affect the taste of the breast milk, as can infections such as mastitis (you can still breastfeed with mastitis, see page 110). Expressed milk, which is frozen and then thawed, may have a slightly different flavor, too.

———

## IS THE BREAST FULLY EMPTIED AT EACH FEED?

No, breast milk is made continually; you don't need to wait for a breast to "fill up." The more milk is drained, the more will be produced. On average, a baby drinks about two-thirds of the milk available at each feed. As your baby feeds from your breast, they will reach the hind milk, which is higher in fat. Don't overthink whether your baby is getting both foremilk and hindmilk; just feed them when they want to be fed and they will get the milk they need.

———

## HOW CAN I MAKE MORE MILK?

Most people will be able to produce enough milk for their babies. During a growth spurt a baby will feed longer or more frequently, meaning you will produce more milk. Signs that your baby is receiving enough milk include: hearing and seeing them swallow; rounded cheeks when they are feeding; they are content and relaxed after feeds; they produce wet and dirty diapers; and they are gaining weight. If you are concerned that you are not making enough milk, then feeding more often can stimulate more milk production, as can expressing or pumping between feeds. Make sure that you are eating and drinking enough, and getting rest—take care of yourself. Some herbs and supplements are marketed as helping increase breast-milk supply, such as fennel, fenugreek, and milk thistle; however, there is not clear evidence that these are effective. If you have a condition such as hypothyroidism, this can impact your milk production, so talk to your doctor.

———

## CAN THERE BE TOO MUCH MILK?

Yes, oversupply can occur. The rush of the let-down reflex can be so strong that it releases more milk than baby can drink in that moment. A sign is the baby coughing or coming off the breast. Holding your baby more upright as you feed and burping more frequently can help your baby cope with the milk flow. Avoid pumping in addition to feeds, as this will encourage more supply. Oversupply can be related to various medications such as antipsychotic drugs, so discuss with your doctor.

———

# how to breastfeed

Learning how to breastfeed your baby takes time, so be patient with yourself. Your baby might take a while to get the hang of it, too.

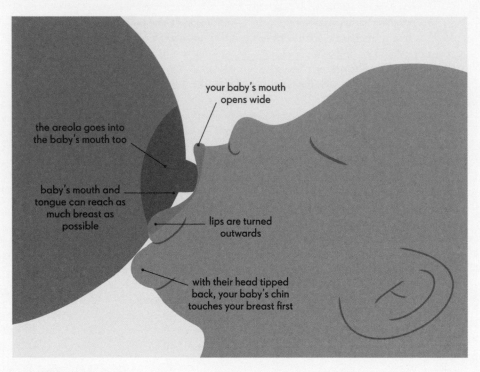

**HOW TO LATCH ON**

# Breastfeeding: step by step

First of all, find a comfortable place to sit or lie down; feeding your baby can take some time. You might want some water to drink, and pillows or cushions nearby to help support baby.

**01.** Hold your baby close to your body so their nose is at the level of your nipple. Support the rest of their body so that it is level and facing toward you, and they aren't twisting their head. Using pillows or cushions to support baby may be helpful.

**02.** Let your nipple brush against their top lip; the baby should then open their mouth wide. As the baby's mouth opens, their chin will come up to touch the breast, and as much of the breast as possible (including the areola) should go into the mouth. This gives a more comfortable latch for both of you. If a baby sucks only on the nipple it will cause pain. Look at the baby's lips: more of the areola should be seen above the baby's top lip than the bottom lip (this also depends on the size of the areola). Baby's lips should be turned outward, rather like a fish.

**03.** Feeding should not be painful, though you may be aware of a tugging sensation inside your breast.

**04.** If you feel that your baby hasn't latched properly, or feeding feels painful, you can remove baby from the breast and start again. Inserting a clean finger into the corner of their mouth can help break the suction so you can remove your breast comfortably.

**05.** Look at your nipple after feeding; it should look similar to normal, not flat, pinched, distorted, or white.

**06.** Lots of people need help when they start breastfeeding, so do ask for help, If you would like it. You can ask your doctor or nurse, and there may also be a lactation advisor in your local community and peer-support groups.

---

**• FINDING A COMFORTABLE POSITION •**

Try different positions to get comfortable during feeding.

**Cradle hold:** the baby rests against your forearm as they feed across your body.

**Football hold:** the baby is held by your side, under your arm (like a football).

**Lying down:** you lie side-by-side.

---

# Breastfeeding with flat or inverted nipples

It is possible to breastfeed with flat or inverted nipples; it is not nipple-feeding after all, but breastfeeding. To latch on well, your baby needs to take in the areola, and not just the nipple, so don't worry. You may find that pregnancy and breastfeeding itself changes the shape of your nipples so that it gets easier over time. It may take more practice to get a good latch, with inverted or flat nipples (see page 32). The following tips can help:

• Roll your nipple between your thumb and fingers to try to get it to stick out. This may depend on the grade of nipple inversion (see page 37).

• Place something cold like a damp cloth over the areola and nipple briefly, to make it erect and stand out.

• Use two fingers in a V-shape, or your thumb and forefinger in a C-shape, and push under the areola and nipple, compressing the breast tissue against your chest, to push the nipple further outward.

• Use a breast pump to pull the nipple outward, or try expressing by hand.

• Wearing a nipple former in the last few weeks of pregnancy may be helpful. These are flexible silicone disks which apply slight pressure to gently pull the nipple outward. Start with a short amount of time and increase gradually. Check with your antenatal team before using.

• Flat or inverted nipples can retract between feeds, so you might need to repeat these techniques again. Dry your nipples before they retract as dampness may increase the likelihood of infection.

• As before (see page 101), do ask for help and support if you would like it.

# Breastfeeding with pierced nipples

It is important to remove nipple jewelery when breastfeeding because it could graze your baby's mouth, or be a choking hazard. Ideally, a nipple piercing would be removed for the entire time of breastfeeding, even though there is a risk that the piercing will close. However, some people choose to remove the jewelery before each feed and replace it afterward. If you are replacing the jewelery each time, make sure it is clean and dry before reinsertion.

If you've removed your jewelery to feed, milk may leak from the piercing holes, causing a faster flow of milk. Although it is safe to feed if you have a nipple piercing, it is recommended that you don't have a new nipple piercing while pregnant or feeding. It is a good idea to wait at least three months after finishing feeding, or have the piercing at least a year before pregnancy. This is because a nipple piercing can take many months to heal, and infections are more likely during the healing process.

( . )( . )

## Breastfeeding and breast surgery

It depends on the surgery you have, but for most breast or nipple surgeries you will still have glandular tissue, so are likely to be able to make milk (though there may be damage to some of the ducts and nerves).

If you have had breast reduction (see page 190, breast hypoplasia) there may not be enough glandular tissue for exclusive breastfeeding but you can certainly try—everyone is unique. Depending on how much tissue is left you may be able to exclusively feed or may need to use supplementary feeding with formula.

If you have had breast augmentation—for example, implants—those which sit below the muscle wall affect milk production less than those placed above the muscle wall. This is all usually discussed before surgery.

Surgery which has involved detaching and replacing the areola and nipple, which has produced scars around the areola, can lead to reduced milk supply because scar tissue may affect the delivery of milk. Over time, the ducts and nerves may form new pathways which can improve milk supply.

A mastectomy on one side means that all the glandular breast tissue has been removed, so it will not be possible to feed on that side, but the other breast can be used for feeding and can make sufficient milk to meet the baby's needs. If you have had a partial mastectomy and/or radiotherapy, this can lead to reduced milk production on that side, but between your two breasts you can produce enough milk to feed your baby.

( . ) ( . )

# breastfeeding

## CAN YOU BREAST-/CHESTFEED IF YOU HAVE NOT GIVEN BIRTH?

It is possible to induce lactation if you haven't given birth. This may involve medication as well as breast stimulation with a breast pump. Milk supplementation (with donated breast milk or formula) may also be needed. The terms breastfeeding and chestfeeding have been used in this book to ensure inclusivity of everyone who may be involved in feeding. The term breastfeeding has been used more frequently, to reflect the most common scenario.

———

## CAN TRANS AND NONBINARY PEOPLE CHESTFEED?

Yes. The hormones required for breast-milk production are released from the pituitary gland. Hormonal medication and breast stimulation may be required, and milk supplementation (see above) is often also needed. If you have had top surgery there may still be enough tissue left to produce some milk. Please be aware that binding (see page 82) during chestfeeding may increase the risk of mastitis. If you are taking testosterone, discuss with your health-care provider if you are considering chestfeeding, as testosterone can interact with the hormone prolactin. It is important to recognize that breast- and chestfeeding can impact gender dysphoria and mental health, so seek support if you need to. Irrespective of whether or not you choose to breast- or chestfeed, skin-to-skin contact helps bonding and can help the baby regulate their temperature and breathing.

———

## HOW DO I STOP PRODUCING MILK?

Milk production is related to supply and demand (see page 97), meaning it will decrease if you feed less often, and as your baby is weaned. If you decide to reduce feeding, it can be a good idea to drop one feed at a time and let your breasts adjust, to avoid getting breast engorgement, blocked milk ducts, or leaking breasts. If you choose not to breastfeed after delivery, you will stop producing milk naturally as it isn't being used. After a pregnancy loss, medication such as cabergoline can be given to block the production of prolactin and stop milk production.

———

# breastfeeding concerns

Breastfeeding is a learning process for both you and baby. Here are some of the most common conditions which you may experience. You can ask for lactation support, or talk to a midwife, nurse, or doctor.

## Sore nipples

**CONDITION** Sore or cracked nipples occur commonly during breastfeeding. The pain is different from the tingling sensations some feel when the let-down reflex occurs (see page 97)—and from the mild tugging sensation as the baby suckles—and the pain does not resolve after feeding. For some people, without support, the pain may cause them to discontinue breastfeeding. Sore nipples are more common at the start of breastfeeding and are usually temporary as the nipples get used to feeding.

If, however, the pain is severe or doesn't go away, then there may be nipple injury which is often due to a poor latch (see page 100) or setting the breast pump too high. If you are having significant discomfort with breastfeeding, you should discuss your symptoms with your lactation consultant or doctor.

**TREATMENT** Allow sore nipples to heal by letting them air-dry after feeding. Use a lanolin- or petroleum-based nipple ointment, or rub in expressed milk after each feed. You can also wear a cotton or natural-fiber bra to help keep moisture away from the breast between feeds. Apply a cold compress such as a washcloth to your sore nipples to reduce pain, and you can take over-the-counter painkillers.

Make sure you see a lactation consultant or your doctor to ensure that your baby has a good latch (see page 100). If you think that your baby has attached incorrectly, remove them from the breast and start again (see page 101). If your baby has an issue which is preventing good latching technique, such as a tongue tie (which restricts the movement of the tongue), you may be referred for treatment to release the frenulum, which typically helps correct the painful latch.

If you have nipple pain, do not avoid feeding or give shorter feeds to try to "rest" your nipples, because this could negatively affect milk supply.

## Engorgement

**CONDITION** Breast engorgement is common when milk starts to be produced, or "comes in," as the colostrum changes to transitional milk (see page 96). It tends to occur about three days after giving birth and affects about two-thirds of people after delivery. The breasts become enlarged, hard to the touch, feel tender, and even wearing a bra can be painful. Nipples can become flat for a period of time.

Engorgement typically lasts a few days as the milk comes in, although it can occur at other times, too; for example: when your baby goes through a growth spurt and starts drinking more milk, thus stimulating your body to upregulate milk production; when your baby starts to sleep more or wait longer between feeds; or when you are trying to discontinue breastfeeding. Engorgement is often worst for the first pregnancy, but don't worry if you don't have this experience—the milk is still being produced. Engorgement can also be more common or severe if you have breast implants (see page 187).

**TREATMENT** Wearing a really comfortable and supportive bra (see page 92) can be helpful, but you don't want anything too tight that causes compression. Using something to cool the feeling of heat and swelling, such as a cool compress in between or before feeds, can be soothing.

Feeding regularly is helpful to reduce engorgement, and you can also hand express or massage your breasts (especially when in a warm bath or shower) to remove just a small amount of milk to ease the pressure if necessary. Be aware that doing this too frequently may cause your breasts to become engorged again.

Although it isn't clear how or why, placing cold cabbage leaves over your breasts can help. Properties of the cabbage may have an anti-inflammatory effect on breast tissue. Keep the cabbage in the fridge to keep it cold. Cover the breasts with the cabbage leaves for 20-30 minutes. If you like, remove the thick, central vein of the leaves so you have a slit for your nipples. Or you can wear the leaves inside a bra. When your breasts have self-regulated to decrease the engorgement you should stop using cabbage leaves because they may impact your milk supply. Some people use cabbage leaves to prevent milk from coming in, decrease supply, or discontinue breastfeeding altogether.

## Leaking breasts

**CONDITION** When you start breastfeeding, leaking is common. When your baby is feeding on one side, the let-down reflex will affect both sides so the other nipple will leak. You can collect this milk for use later, if you like. Leaking can also occur when you think about your baby—for example, if you hear them crying.

**TREATMENT** Leaking tends to occur less frequently once breastfeeding is established—after about six weeks—but it can still occur after that point. Wearing nipple or nursing pads inside your bra can help prevent you feeling damp and avoid any visible leakage through clothing. Avoid pumping or expressing when you are leaking as this will encourage further milk production.

## Uneven breasts

**CONDITION** Some babies seem to prefer feeding on one breast or the other, or you may notice that one breast may look bigger or smaller than the other. Once you stop breastfeeding this unevenness should improve. Remember, it is normal for your breasts to be asymmetrical (see page 31).

**TREATMENT** Try to offer your baby the less-preferred side first, or pump on that side to encourage production. If you continue to feed more on one side, production in that breast will continue to increase, exacerbating the problem.

( . ) ( . )

## Blocked milk duct

**CONDITION** Milk ducts can become blocked or clogged during breastfeeding, meaning that the milk from the gland behind cannot be drained during feeding, leading to a red and tender lump in your breast. It can, but does not always, lead to an infection of the breast (mastitis—see page 110).

**TREATMENT** To treat a blocked milk duct, keep feeding; this encourages the milk to flow and can help unblock the duct. Using a warm compress before feeds and massaging the breast during feeds can also be helpful. You can also try using an electric toothbrush or vibrator to massage the breast. If more of the breast becomes red and painful, or you develop fevers and chills, please see your doctor, as this could indicate mastitis (see page 110).

## Milk blister

**CONDITION** A milk blister (or "bleb") occurs when a nipple pore becomes blocked—often related to a blocked milk duct—leading to a small white or yellowish spot on the nipple. The surrounding skin may be red or inflamed. Feeding on the affected side may cause nipple pain.

**TREATMENT** Milk blisters generally resolve within a couple days. Applying a warm compress can help, as can continuing to feed on that side as suckling may help open the blister (this won't harm the baby). You can also wear a nipple pad soaked in a little olive oil to try and soften the skin on top of the blister to help it open. You could also try compressing the nipple just behind the blockage to try and express out the blockage. If it does not resolve or gets worse, please see your doctor.

## Nipple vasospasm

**CONDITION** Vasospasm is the term for when the tiny blood vessels supplying the nipple go into spasm and constrict the blood flow to the nipple. Nipple vasospasm is more common in people who have Raynaud's phenomenon (see page 131). The nipple may turn blue or white, go numb, and then turn pink or red and become painful as blood flow returns.

**TREATMENT** To try and prevent vasospasm, place a warm compress on your breast prior to feeding. Keep your body warm; if you get cold during breastfeeding, wear a scarf or cardigan to wrap up as much as possible, especially at night. Cover your nipple soon after feeding. To relieve discomfort, use a warm compress. If you have any concerns, please see your doctor.

## Thrush

**CONDITION** Thrush is a fungal infection that can affect your nipples during breastfeeding. Your nipples may be browner or darker than usual, or pinker or redder than usual if you have light skin, or you might notice some crusting. You may feel a burning sensation or a deeper shooting pain in the breast during or after feeding. Or you may not have any symptoms but might notice it on baby instead, with white or yellow patches in your baby's mouth. These patches are different to milk residue in that they cannot be wiped away.

**TREATMENT** If you or your baby has symptoms of thrush then your doctor may recommend treatment with an antifungal medication in the form of a gel, liquid, or cream applied to the nipples. Depending on which product is used, you may or may not have to wash it off before feeding.

( . ) ( . )

## Mastitis

**CONDITION** Mastitis is an infection of the breast itself and is relatively common during breastfeeding; estimates vary but it may affect up to 1 in 5 people within the first six weeks or so after delivery. Mastitis can occur after a blocked duct, or from sore or cracked nipples.

Mastitis leads to an area of the breast becoming red, hot, and tender, and can be extremely uncomfortable. It generally affects a wedge-shaped area of the breast, which may look swollen. You may also have a fever, chills, or generalized body aches. If you have symptoms of mastitis, contact your doctor as soon as possible.

**TREATMENT** Mastitis is treated with antibiotics, which your doctor will prescribe for you. You can also use over-the-counter painkillers for pain relief. Warm compresses are also helpful.

You should continue to breastfeed on the affected side as this can help prevent further engorgement. There is no risk to the baby from continued breastfeeding, or from the antibiotics used to treat mastitis. If feeding from the affected side is too painful, then expressing with pump or your hands may be easier.

Your baby may refuse to feed from the affected side due to changes in how the milk may taste. In this situation, you can continue to feed from the other, unaffected, breast.

If your symptoms don't improve within 24-48 hours, or persist after you have finished your antibiotics, you should contact your doctor as this may indicate that a pocket of infection, or an abscess, may have formed.

## Galactocele

**CONDITION** A galactocele is a benign cyst in the breast which is filled with breast milk. Galactoceles present as smooth, round swellings or a lump in the breast. Unlike blocked milk ducts (see page 108) these are not painful and the overlying skin is not red or hot. Pressing the lump can cause milk to flow from the nipple.

**TREATMENT** If you find a lump in your breast, you should be assessed by a doctor. You may be offered imaging investigations such as an ultrasound. Galactoceles tend to go away spontaneously after stopping feeding, or can be drained.

## Breast abscess

**CONDITION** An abscess is a localized collection of pus, and is sometimes linked with mastitis that has been incompletely treated. If you have an abscess you are likely to feel unwell. You may also have an area of the breast which is red, hot, swollen, and painful. There may also be some discharge of pus. Unlike with mastitis, if you have a breast abscess you should avoid breastfeeding from the affected side.

**TREATMENT** If you are experiencing symptoms of a breast abscess, you will need to be evaluated by your doctor. Treatment involves drainage of the abscess and often antibiotics. Although you should not breastfeed from a breast with an abscess, it is helpful to express the milk from that breast, to avoid or treat engorgement and ensure that milk supply is not affected. You should then discard this milk.

### • SELF-CHECKS •

It is possible to develop breast cancer during pregnancy and breastfeeding, so if you feel a new lump or have any concerns about changes in your breasts, please contact your doctor to schedule an exam.

(.)(.)

# breastfeeding myths

### "BREASTFEEDING IS NATURAL SO IT MUST BE EASY"

It may be natural but it isn't always easy! It can take a while for both you and baby to learn how to feed easily, so do ask for the support you need.

———

### "BREASTFEEDING IS PAINFUL"

Your breasts may feel tender for the first few days but feeding should not be painful. If it is painful, ask for support from a lactation advisor.

———

### "IF YOUR BABY LOSES WEIGHT IN THE FIRST FEW DAYS YOU AREN'T MAKING ENOUGH MILK"

Colostrum is produced in the first few days until the transitional milk comes in, and it is normal for babies to lose a little weight until then. Generally, babies lose less than 10 percent of their birth weight and then start gaining again as they start to drink milk. Your doctor will weigh your baby and advise on early feeding.

———

### "BREASTFEEDING IS ESSENTIAL FOR BONDING"

While breastfeeding releases hormones which can help with feelings of connection and bonding (see page 97), rest assured that you can bond with your baby however they are fed.

———

### "YOU JUST SIT THERE DURING FEEDING, SO YOU SHOULDN'T BE TIRED"

Breastfeeding takes energy—you are making the milk to sustain a baby, and this means that, on average, you need an extra 500 calories per day while feeding. Add in lack of sleep and daily baby care, and of course you may be tired.

———

### "IF YOU WANT TO BREASTFEED YOU MUST NEVER GIVE A BOTTLE"

Many people will choose to mix feed, or supplement breastfeeding with expressed milk or formula. "Fed is best," and you might find that the occasional bottle can give you time to rest and recover before the next breastfeed.

———

### "A CRYING BABY MEANS YOU AREN'T PRODUCING ENOUGH BREAST MILK"

Babies cry for many reasons, not only hunger. If your baby is urinating well, seems satisfied after feeding, and is gaining weight, they are likely to be getting enough milk. If you are concerned, speak to your doctor.

———

### "SMALL BREASTS ARE NOT ABLE TO PRODUCE ENOUGH MILK"

No, the ability to produce milk is not related to the size of your breasts (see page 98).

———

### "BREASTFEEDING WILL MAKE YOUR BREASTS SAG"

It hasn't been shown that breastfeeding causes changes to breast volume, shape, or sagging. However, the changes in your breasts during pregnancy (see page 88) as well as weight changes, may contribute to the ligaments in the breasts being stretched, which may impact on their overall shape.

———

### "YOU MUST NEVER DRINK ALCOHOL WHILE BREASTFEEDING"

An occasional drink (1-2 units approximately once a week) is unlikely to cause harm. If you do drink alcohol, wait a few hours before feeding your baby. This is because alcohol passes into the breast milk about 30-90 minutes after drinking, and then takes a few hours to clear your body, depending on how much alcohol you have drunk.

———

# 06

# Menopause and beyond

# breast changes in midlife

Fluctuating and falling hormone levels in midlife can lead to breast symptoms and changes. You may also be offered breast-screening, let's look at what's involved.

## During perimenopause

Perimenopause (or menopause transition) is the period of time leading up to your last period, during which time your periods may be regular or irregular. This means that your breasts can be irregularly exposed to estrogen and progesterone and may become tender or sore.

Perimenopause can start in your mid-40s, which might be sooner than you expect, so if this is you, it is worth reading this section so you know what to expect. If you go through a premature or early menopause you may notice these changes earlier. Menopause is defined as the last period, but you can't tell until you haven't had a period for 12 months. After this point you are considered to be post-menopausal.

## After menopause

Falling levels of estrogen and progesterone affect the whole body, including your breasts. The glandular tissue, which is involved in milk production, shrinks and the breasts become less dense, being mostly made up of fatty tissue. In addition, the levels of collagen and elastin, which give strength, support, and elasticity to the skin, decrease. This can lead to dry and dehydrated-looking skin and the appearance of wrinkles. The combination of these changes may lead to the breasts looking smaller, lower down, elongated, or flattened. The nipples may point downward and you may notice more space between the breasts. You may also notice the development of, or more, hairs around the areolae; these are normal and due to hormonal changes.

### • HRT •

Hormone replacement therapy (HRT) can be used to treat symptoms of perimenopause and menopause, such as hot flashes or mood changes, by increasing the levels of hormones. HRT contains estrogen, and if you still have your womb, then you will also receive progesterone. Some people will also receive the hormone testosterone. Breast soreness and tenderness can be a side-effect of the estrogen or the progesterone component of HRT. Although it can take a few months, breast soreness tends to settle with time.

## Breast cysts

As you get older, the risk of developing benign cysts and other growths in the breasts increases. A cyst is a fluid-filled sac. We don't yet know why they form. Breast cysts are benign and don't increase your likelihood of developing cancer. Often, the lumps you feel in your breasts are cysts. Cysts can make it more difficult for you to feel other new lumps, so be sure to know your breasts well. If you notice any changes during your self-examinations (see pages 39–43) please see your doctor.

## Breast sagging

Sagging, or ptosis, is common in midlife and later. Just like the rest of you, your breasts change in appearance over time and while sagging is more common after menopause, it can occur at any age. Ptosis can occur in breasts of all sizes, though it is more likely in larger breasts. Breast ptosis can vary in appearance— for example, the breasts may seem to be fuller at the bottom and less full at the top than previously; they may seem to sit lower down on the chest; or the nipples can change from pointing forward to pointing downward.

### GRADES OF SAGGING

This refers to the position of the nipple in relation to the inframammary fold (the line underneath the breast). Knowing the grade of breast sagging may be useful if you are considering cosmetic surgery. Breasts and bodies change and there should be no judgment associated with that.

 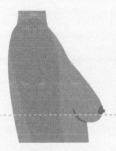 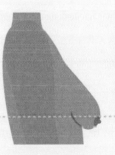 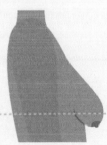

**NO PTOSIS** the nipple is above the inframammary fold.

**GRADE 1** the nipple is in line with the fold.

**GRADE 2** the nipple is below the fold but is not the lowest part of the breast.

**GRADE 3** the nipple is pointing downward at the lowest part of the breast.

(.)(.)

# breast sagging (ptosis)

## WHY DOES BREAST SAGGING OCCUR?

Breast support comes from the skin, ligaments, and connective tissue. Over time, changes in the skin mean that it becomes less elastic and less able to support the weight of the breasts against gravity. Hormone changes during the menstrual cycle, pregnancy, breastfeeding, perimenopause, and menopause can all contribute to breast changes. Weight loss and weight gain also affect breast sagging. Smoking can damage your skin, including the skin on your breasts. Not wearing breast support, in particular during exercise, can also stretch the ligaments and skin and may contribute to sagging of the breasts.

———

## CAN SAGGING BREASTS BE PREVENTED?

There are lots of factors which lead to breast ptosis, some of which, like aging and gravity, are not in your control. However, stopping, or not starting smoking may help; smoking damages collagen and elastin in the skin, weakening it as well as narrowing the blood vessels which supply the skin. Trying to maintain a healthy weight and wearing well-fitting bras are also helpful.

———

## CAN SAGGING BREASTS BE TREATED?

Wearing a well-fitting bra (see page 63) can support the breasts and help with shape and appearance. The breasts themselves don't contain muscle, so there are no exercises to increase the size of the breasts. However, there are muscles on the chest wall underneath the breasts, so exercising the chest and shoulder muscles may help to slightly lift the breasts. Surgery to lift the breasts (mastopexy) can also be used (see chapter 9).

———

# breast screening

In many countries, breast-cancer screening starts around the
time of perimenopause and menopause.

Breast-screening programs tend to start at around age 50, or earlier in some countries. This is because the risk of breast cancer increases with age, with rates of breast cancer in the UK being 11.5 per 100,000 between the ages of 25 and 29, compared with 285.5 per 100,000 between the ages of 55 and 59.

The rate of breast-cancer diagnoses rises with age, but seems to plateau for a few years after the age of 50 (in the UK) before rising again; this is likely to be due to the onset of screening programs around this time.

Put simply, breast screening can save lives. In the US, it is estimated that around half a million deaths from breast cancer have been prevented between 1989 and 2019, due to the breast-screening program and better treatment. In the UK, screening prevents around 1,300 deaths from breast cancer every year. An NHS report covering 2019-2020 found more than 17,500 cases of breast cancer were discovered during breast screening. Results of studies have been varied, but an independent review in 2012 in the UK concluded that for every 10,000 women who attend screening between the ages of 50 and 70 years old, 43 lives are saved. An international review concluded that for every 2,000 women screened for a decade, one life would be saved.

The World Health Organization recognizes that countries, health-care systems, and screening programs differ around the world, and as such, the WHO recommendations differ according to the resources available. A breast-screening program is not the only puzzle piece in prevention; it is still very important to remain breast aware and to check your breasts regularly (see pages 39-43). Breast cancer can develop between mammograms, or before you have your first mammogram, so regular checking is important. If you develop any symptoms or changes in your breasts, please see your doctor—do not wait until your regular breast screening.

---

## • HIGH-RISK GROUPS •

The information in this section is aimed at people who are not at an increased risk of breast cancer. For information about screening for those at higher risk of developing breast cancer, please see page 157.

## Breast-screening programs around the world

In the UK, the NHS breast-screening program starts at age 50. Women between the ages of 50 and 70 are invited for breast screening with a mammogram, every three years. After the age of 70 you are still entitled to a mammogram every 3 years, but will not routinely be invited for one; instead, contact your local screening unit to arrange an appointment.

In the US, organizations such as the American Cancer Society, the US Preventative Services Task Force, and the American College of Obstetricians and Gynecologists have different recommendations and guidelines on the frequency and timing of breast screening.

However, there is an agreement that people should be offered breast screening between the ages of 50 and 74, though some advise annual screening, and others biennial. Screening between the ages of 40 and 49 may be offered, and often involves a discussion of potential risks and benefits (see page 124).

The Australian breast-screening program (BreastScreen Australia) involves a biennial mammogram between the ages of 50 and 74. Between the ages of 40 and 49, and after the age of 75, you can still have mammograms but won't automatically be invited for them.

In Germany, mammograms are offered every two years between 50 and 69 and in France, every two years between 50 and 74.

# Screening with mammography has helped reduce breast cancer mortality by 40 percent; this means the risk of dying from breast cancer is nearly halved.

# your breast screening appointment

Your appointment is scheduled. Let's look at how to prepare, what will happen, and address any concerns you might have.

## What is a mammogram?

The test for breast screening involves a mammogram, which is a low-dose X-ray of the breasts. Breast lumps may contain denser (more solid) tissue which blocks more X-rays and can be seen on the image. You might be interested to know that lower doses of radiation can be used now than in previous years, as the images created are digital, which requires less radiation than film to create high-quality images.

## How to prepare

On the day of your screening please do not use deodorant or talcum powder, because these may show up on the mammogram, making it difficult to read. You will be asked to remove your clothes on your top half, down to the waist, including your bra (if you wear one), and put on a gown. You may feel more comfortable to go to the appointment in separates, so that even when your top is off you are still wearing trousers or a skirt. You will also be asked to remove jewelery in the area such as nipple piercings or necklaces. The appointment itself takes approximately half an hour.

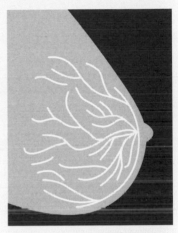

Mammogram showing normal breast tissue

Mammogram showing a dense lump in the breast

(.)(.)

## Breast positioning

The radiographer will show you the correct position to stand in, in front of the mammogram machine. The breasts need to be positioned between two X-ray plates.

Two X-ray views are taken of each breast—one from above, and the second an angled view looking into the armpit and diagonally across the breast. This maximizes the amount of tissue captured in the X-ray images of your breast so as much information is gathered as possible. You will be asked to turn your body, or hold your arm up or out, to get into the best positions. You then stay still while the X-ray is taken.

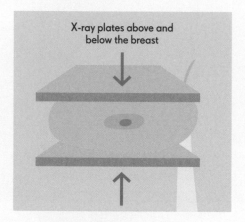

**X-ray plates above and below the breast**

HORIZONTAL X-RAY POSITION

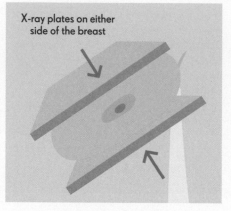

**X-ray plates on either side of the breast**

DIAGONAL X-RAY POSITION

---

**Q: I HAVE BEEN TOLD I HAVE DENSE BREASTS: WHAT DOES THAT MEAN?**

—————————— ( . ) ( . ) ——————————

**A:** *Breasts are made up of glandular, fibrous, and fatty tissue. "Dense" breasts have proportionately more glandular and fibrous tissue than fatty tissue. It is not possible to tell if your breasts are dense by examining by look or feel, and it first tends to get identified on a mammogram. Dense breasts and cancer are both seen as white areas on a mammogram, and having dense breasts increases the risk of developing breast cancer; therefore, if you have dense breasts, you may be advised to have additional imaging or investigations.*

( . ) ( . )

## Breast screening for trans and nonbinary people

In the UK, breast screening is available to everyone over the age of 50 who has breasts due to naturally-occurring estrogen or due to estrogen hormone therapy. This includes trans men and nonbinary people assigned female at birth, and trans women and nonbinary people assigned male at birth who are taking estrogen.

Remember that even after top surgery, there may be some breast tissue remaining, so it is still important to attend screening. If you wear a breast or chest binder (see page 82), you will be asked to remove this during your screening appointment.

## When will I get my results?

Do ask your radiographer when you are likely to get your results, as this can vary between screening programs. In the US, you should receive your results in one to two weeks. The use of artificial intelligence in image-processing is being researched as a way to speed up delivering the results. Both you and your doctor will be sent your screening results. If you have a normal result, no further action is required, and you will be asked to arrange breast screening after the screening interval according to where you live and your age (see page 119). However, if you develop any symptoms or breast changes before then, please don't wait for the next mammogram—contact your doctor.

Occasionally the mammogram will not produce a clear image and you may be recalled to repeat the screening or referred to a specialist.

## What if my results show an abnormality?

About 4 in every 100 women screened in the UK receive an abnormal mammogram result and will be called back. In the US about 10 in every 100 get called back. Having an abnormal mammogram does not always mean that you have breast cancer; in fact, most people who are recalled after breast screening do not have cancer. Of those 10 called back in the US, only about 1 has cancer. Of the 4 called back in the UK, only about 1 has cancer.

You may be asked to return for a magnified mammogram, where a particular area of the breast is looked at more closely, or you may be referred to a breast clinic to see a specialist to decide on the next steps. For more information about what happens in breast clinic after an abnormal mammogram, or finding a breast change, see page 136.

(.)(.)

# breast-screening concerns?

Not every person who is eligible for screening goes to an appointment,
so let's look at any potential concerns you might have.

Approximately 7 out of 10 women in the UK and US regularly attend breast screening. There are lots of reasons why someone may not attend, from worries about the process of the screening test to worries about the results. Let's answer some common concerns about this potentially life-saving screening.

All tests and treatments in medicine (like most decisions in life, in general) have pros and cons, and the potential risks and benefits need to be weighed up. In the balance, the potential risks of breast-cancer screening are outweighed by the benefits of earlier detection and saving lives. The main advantage of breast screening is that it can pick up breast cancer at a stage before symptoms would otherwise be discovered, so potentially life-saving treatment can begin sooner. Screening saves lives.

## Does it hurt?

In order to take the mammogram, the breasts need to be positioned between two X-ray plates. This may involve compressing the breasts slightly which can cause a squeezing sensation, discomfort, or even pain. This discomfort tends to last just a few seconds while the X-ray is taken until the pressure is released, but some women will notice it feels worse before their period is due (the average age for menopause is 51, meaning that some people will be eligible for the screening program and still be having periods). Others may notice that the breast discomfort lasts a few days. The X-raying process only takes a few minutes, and then you will be on your way.

---

### • BREAST SCREENING AND BREAST IMPLANTS •

If you have breast implants, please inform the screening unit. This is because some of the breast tissue might be hidden by the implant, so you may need further X-ray views. The screening does not check the implant itself, so if you have concerns about your implants, discuss with your doctor.

---

(.)(.)

## Anxiety

Going for a mammogram and waiting for results can themselves cause anxiety. If you have an inconclusive result, or have to be recalled for further testing, this can increase that anxiety, even though the majority of those recalled will not have cancer. Remember that just 10 people out of every 100 in the UK are recalled after breast screening (see page 123), and in most cases will not have breast cancer.

There is evidence that anxiety may reduce the likelihood of people returning to screening after being recalled. Please be assured that it is still more than likely that you will be fine. Of the 10 women out of every 100 recalled in the US, 90 percent will not have cause for concern, so please do get checked.

## False negatives

Although it is rare, occasionally breast screening may have some false negatives—the result will be negative but the person has early breast cancer, which is missed. This may occur because it is hard to see the cancer on the mammogram and is estimated to occur in 1 in every 2,500 screening tests taken. This is why regular self-examinations (see pages 39-43) are so important, even if you have had screening.

## Over-diagnosis

This refers not to breast cancer which would later develop symptoms, but to breast cancer which would not have been picked up without breast screening and would not later cause concern. It is possible to have a cancer which is not life-threatening and may never lead to symptoms.

This means that some people will be offered treatment that they potentially do not need. However, research in the UK has shown that the risk of over-diagnosis is small, estimated at just under 3 per 1,000. This corresponds to approximately 3.7 percent of all cancers being picked up at the screening program being over-diagnosed. This is far lower than previously thought; previous numbers were anywhere from 5-30 percent.

It is not currently possible to predict which cancer may progress extremely slowly and not cause concern, and which will develop more aggressively, so if a cancer is detected at breast screening, treatment will be offered.

## Radiation

A mammogram is a form of X-ray and therefore involves radiation. The amount of radiation used is small, so the risk from it is small. A review in 2014 stated that for 10,000 mammograms there are between 1 and 10 people who develop breast cancer related to the radiation. In fact, the amount of radiation used is about the same as seven weeks of background radiation absorbed from simply living.

# 07

# When things go wrong

# breast pain and discomfort

Many of us will experience breast pain (mastalgia) or discomfort at some point in our lives. Let's look at what is going on.

A chapter called "When things go wrong" doesn't sound reassuring, but there are lots of symptoms and conditions which can affect the breast and it is important to be aware of them, so that you can get help, if needed. Many patients are concerned that any breast symptom means breast cancer, but benign conditions are common (though, of course, any symptom needs to be assessed).

Around 7 in 10 women will experience breast pain (mastalgia) or discomfort at some point. For the majority, the pain is mild and manageable, often related to the menstrual cycle. For 1 in 10, breast pain can be moderate to severe and have a significant impact on their lives. Using a calendar to track your pain and your menstrual cycle for a few months may help you identify if your pain is cyclical. Make a note of when the pain comes, your menstrual cycle, how bad the pain is (perhaps out of 10), or when it impacted your activities, sleep, or lifestyle.

When we have unexplained breast pain we often jump to anxious, worst-case scenario conclusions. Assuming that your pain is a symptom of breast cancer is very common. Although this is possible, breast pain alone is not a common presentation of breast cancer— generally it presents with a painless lump or thickened area (see page 160). Pain affecting both breasts is unlikely to be related to cancer. However, if you have unusual breast pain, or any concerns, then see your doctor.

## Cyclical breast pain

Cyclical breast pain is extremely common and is related to the hormone changes of the menstrual cycle and how these impact the breasts (see page 70). It can occur at any age from as soon as periods start, to when the menstrual cycle ends at the menopause. It is often worse during the younger years and then again during perimenopause (see page 116). It may be that some people have breast tissue which is more sensitive to these hormone changes than others.

The severity of the pain and discomfort varies between women, but also from month to month. For many women it is mild, but for others it has an impact on their lives; for example, they may avoid physical activity and sports during that time, avoid hugging, and it may impact their libido or make sex painful. All pain can interrupt sleep (especially if you are used to sleeping on your front) which can lead to fatigue and impact your mood.

Describing the pain and when it may be worse or better is not done to scare you—after all, everyone is likely to have different symptoms— but to help you understand your body, and therefore understand what might help. Keep reading—tips to help cyclical breast pain are on the next page.

Generally, the pain is worst in the few days leading up to the period starting, but can start earlier, after ovulation, up to two weeks before the period (see chart on page 70). It tends to improve after the period starts, generally resolving by the time the period has finished.

The pain and tenderness can affect both breasts, though you may notice that the severity is different in each breast. The pain can occur anywhere on the breast but is often worst on the upper, outer part of the breast or extend up into the underarm. You may also notice changes in the breasts during this time—they may feel more swollen and lumpy all over—though these changes tend to resolve after your period begins.

## WILL IT GO AWAY?

Having cyclical mastalgia does not mean that you are always going to have breast pain each month for the rest of the time that you have menstrual cycles; it may disappear or come and go. In fact, in 20-30 percent of women, the pain improves within three months or so. However, for 6 out of 10 of these women, the pain may reappear in the next couple of years. As yet, the cause of the fluctuating nature of the pain is not known, it may be related to hormone changes and fluctuations throughout your life.

## TREATMENT

If your discomfort is mild, then no treatment is required. The first thing to do is to make sure you are wearing supportive, well-fitting bras (see page 63). If you have pain at night, sleeping in a soft bra without underwire might help. You may choose to wear a soft bra at night, and a more supportive one during the day.

---

### • WILL SUPPLEMENTS HELP? •

There is some, but not significant, evidence that taking 25-30 g of flaxseed per day may help alleviate cyclical mastalgia. You could try adding ground flaxseed to smoothies, cereals, or yogurt, but this is not routinely suggested as a treatment for mastalgia (it can, however, help with constipation).

Some women find vitamin E, vitamin B6, or evening primrose oil help with breast pain, but there is not robust data to prove these are beneficial.

Before considering new supplements, make sure you check with your pharmacist or doctor to ensure that the supplements do not interact with any medications you may already take.

---

• Over the counter pain relievers can be used to reduce breast pain and discomfort. Only take the recommended amount as detailed on the packaging.

• Topical non-steroidal anti-inflammatory drugs (NSAIDs) like ibuprofen gel, can be rubbed directly onto the skin. They have been shown to be helpful for cyclical mastalgia.

• Hormonal treatments such as contraceptive pills might help prevent symptoms. See pages 73 and 150.

(.)(.)

- In rare circumstances, medications can be given to either reduce or block estrogen, such as tamoxifen or injections of goserelin. These medications often have menopausal symptoms as side-effects and are generally only used if no other treatments are working and the pain is having a significant impact on your life.

## Non-cyclical breast pain

Breast pain is not always related to the menstrual cycle, and you may be able to figure this out by tracking. It may be intermittent or constant. Unlike cyclical breast pain, which generally affects both breasts, non-cyclical breast pain tends to affect one side only, but not always.

Non-cyclical breast pain could be related to infection such as mastitis (see page 110). Other infections, like shingles, may also cause breast pain (see page 133). If you have ongoing breast pain, see your doctor.

Pregnancy and breastfeeding can also lead to breast and nipple discomfort (see chapter 5).

## Chest-wall pain

Although pain may be felt in the breast, the underlying issue might not be related to the breast tissue itself, but another part of the body. Pain in the muscles of the chest wall, or the rib cage and sternum (breastbone), can be felt in the breasts. Pain from the chest muscles could be because of a poorly fitting bra, as the weight of the breasts hangs away from the chest wall, or due to exercise strain or injury.

Costochondritis is a condition where there is inflammation of one or more of the joints between your ribs and sternum which can lead to pain in the area or breast. This is often worse on movement or deep breathing, or on touching the area. Muscle injury to the pectoralis muscles (pecs) can also lead to breast pain.

Treatment may depend on the cause of the pain, but non-cyclical mastalgia resolves on its own in about half of women. A well-fitting bra is important (see page 63). Stretching and mobility exercises for the back, shoulders, and neck can be helpful if you are experiencing muscular pain. Over-the-counter pain relievers, as well as topical NSAIDs can also be helpful.

Discuss with your pharmacist or doctor whether any medication you are taking could be causing the breast pain, for example, the combined oral contraceptive pill or HRT might have an impact on pain. Antidepressants, such as sertraline, and other medications might also have an impact. You may be advised to wait to see if the pain settles down, or consider switching or stopping the medication.

# Pain in the muscles of the chest wall, or the rib cage and sternum (breastbone), can be felt in the breasts.

# nipple pain and discomfort

Nipples are densely supplied with nerve endings, meaning that they are extremely sensitive to both pleasure and discomfort.

Just as you may get breast tenderness or discomfort in the lead up to and during a period (see page 71), the nipples may also be painful or tender.

Nipples can also be painful due to chafing and irritation from clothes, such as a poorly fitting bra or shirt. Nipple-chafing can occur with repeated friction and motion, such as during long-distance running, leading to pain or bleeding. To avoid this, wear a well-fitting sports bra and consider covering the nipples with nipple guards, shields, bandages, or medical tape. Applying an anti-chafing cream or a thin layer of petroleum jelly may be useful too, though you may need to reapply during a long run.

For more information about nipple pain during breastfeeding, including thrush infections of the nipple, see page 109.

## Skin conditions

The skin of the nipple, areola, and breast can be affected by allergic reactions or skin conditions such as eczema and psoriasis, which may cause itching and irritation. Using cotton bras, or a cotton insert into a bra (for example, cut up a cotton pillowcase) may help if irritation is coming from the bra itself. It is important to get any nipple changes (see page 40) assessed by your doctor.

There isn't evidence that biological washing products worsen eczema; however, some people feel that they do and prefer to use non-bio. Fabric softener makes the laundry soft but is left on the product and can cause irritation.

Washing your skin with an emollient or cleanser which doesn't strip the skin of moisture can be helpful for dry-skin conditions.

## Nipple vasospasm

Vasospasm is the term used when blood vessels constrict and narrow, reducing blood supply to the area. In nipple vasospasm, this affects the nipples, which may turn blue and then white, and cause pain which is often described as a burning, numb sensation. When the area warms up and the blood supply returns to the nipple, the nipple turns pink or red and can be painful. Nipple vasospasm can occur related to breastfeeding (see page 109). The main treatment for nipple vasospasm is keeping your whole body warm and your nipples covered.

Raynaud's disease is a condition which generally affects the extremities—your fingers and toes, and, of course, the nipple is also an extremity. The symptoms are very similar to vasospasm. Raynaud's disease is related to other connective-tissue disorders so discuss with your doctor.

# nipple discharge

The most common cause of nipple discharge is breastfeeding or pregnancy, but it can occur at other times, and in all genders.

Galactorrhea is the medical term for producing milky fluid from the nipple unrelated to pregnancy or feeding. It may be related to your menstrual cycle and be normal for you. Breast stimulation can bring on nipple discharge—for example, during sexual activity, or from prolonged and repetitive friction from clothes. It can also be spontaneous, happening on its own, or be related to your menstrual cycle and be normal for you. Galactorrhea can occur in infants, in particular newborns (see page 29), and in all genders. Nipple discharge may also be associated with other symptoms, such as irregular periods. However, other forms of nipple discharge can occur. If you aren't breastfeeding or pregnant, nipple discharge should be checked out by your doctor.

cord. It can also be induced by medications, drugs, and supplements, such as hormonal contraceptive pills, certain antidepressants, high blood pressure medication, use of opioids, and herbal supplements such as fenugreek and fennel. It can even be related to stress. In males it may be related to testosterone deficiency.

Sometimes no cause is found, which is called idiopathic galactorrhea. One theory is that the glandular tissue of the breast is extra-sensitive to prolactin, meaning that even if the prolactin level is normal, galactorrhea may occur.

## Potential causes

Galactorrhea can be related to having high levels of prolactin, one of the hormones involved in milk production. Prolactin is made in the pituitary gland in the brain, so if a non-malignant (benign) tumor occurs in the pituitary it can cause too much prolactin to be released.

Other causes of galactorrhea include kidney disease, hypothyroidism (an underactive thyroid), or nerve damage to the chest or spinal

### • SEE YOUR DOCTOR •

Male nipple discharge should always be checked by your doctor. If you have nipple discharge unrelated to pregnancy or breastfeeding, then see your doctor. Nipple discharge isn't always milky colored: it could be clear, yellowish-green, or blood-stained.

(.)(.)

# breast rashes

Here we look at the most common types of rash, how to
prevent rashes, and when to contact your doctor.

While most breast rashes are not related to
cancer, it is important that all rashes are
assessed, as a rash can be a sign of breast
cancer. You might also notice that the skin of
the breast, nipple, and/or areola may be itchy,
burning, or tingling. This might be in addition
to discharge from the nipple (see page 132).
Other breast conditions such as mastitis or
a breast abscess can also affect the skin.

## Intertrigo

This is a common type of rash generally found
underneath the breasts, particularly if you have
large or pendulous breasts, or if you sweat a lot.
Intertrigo occurs when skin rubs against skin,
and if the skin is moist it is more likely to occur.
A bacterial or fungal infection can also arise.
Symptoms of intertrigo of the breast include:

• Itching.
• Sore skin.
• Changes in skin color—may be red,
  browner, or darker than usual.
• Skin cracking, sores, or ulcerated areas.
• Odor.

To treat intertrigo, keep the area clean and dry,
wash as normal, then pat the skin dry instead of
rubbing it (or you could use a hairdryer on the
cool setting). If you have obesity or are

overweight, losing weight may also be helpful
to reduce the chance of future rashes. If a fungal
or bacterial infection develops, medication can
be given, generally in the form of an antibiotic
or antifungal cream. Make sure that your bra fits
well and isn't causing chafing; you may find
natural fibers like cotton helpful. Change your
bra daily. Consider using a barrier cream to
reduce future chafing and friction.

## Shingles

If you have ever had chickenpox then it is
possible that you may develop shingles at some
point. Shingles is a reactivation of the virus
which causes chickenpox, which lies dormant in
a part of the nervous system. If it reactivates it
causes symptoms in the skin affected by that
nerve, commonly causing a painful rash. If
this happens in a band across your chest or
abdomen, it can affect the skin over the breasts.
The rash tends to look like small blisters which
then crust over. Antiviral medications can be
prescribed and painkillers can also be used.

## Contact dermatitis

Contact dermatitis is an allergy or a reaction
to an irritant such as a new perfume or body
lotion. The skin can become inflamed and itchy.

(.)(.)

Treatment for contact dermatitis involves avoiding the product causing the rash, and using antihistamine tablets for the itch. Steroid and other creams may also be useful. If you develop symptoms of anaphylaxis, such as facial swelling or difficulties breathing, please seek urgent help.

## Other skin conditions

The breasts are covered in skin, meaning that you can get inflammatory skin conditions on the breast such as eczema and psoriasis. Eczema is a skin condition which leads to very dry skin which can become inflamed, cracked, and very itchy. Psoriasis tends to appear as red patches often covered in scale.

Treatment will depend on the condition but often involve emollients to moisturize and steroid creams or ointments. If you have a new rash on your breasts seek medical advice as it may not be easy to distinguish the cause.

## Inflammatory breast cancer

Inflammatory breast cancer is a rare type of breast cancer but can present with skin changes, thought to be due to the cancer cells blocking the lymphatic vessels to the skin leading to the appearance of inflammation. Symptoms of inflammatory breast cancer may include:

• Breast swelling. If one breast has increased in size, remember some asymmetry between the breasts is normal, but if this changes or becomes more pronounced do get it checked.

• Changes to skin color—often red, pink, or darker areas. Darker areas are more common on melanated skin.
• Changes to skin texture. The skin may have small indents on it, or small ridges, and may resemble the skin of an orange (known as "peau d'orange") or look like cellulite. The skin may also look thicker.
• Itching of the skin, or a feeling of heat or burning sensation.
• Nipple changes such as flattening, inversion, or change in direction.

## Paget's disease

Paget's disease of the breast is a rare form of cancer which affects the skin of the nipple and areola. It may also be associated with another breast cancer in the same breast. Although the skin changes of eczema and Paget's disease may appear similar, Paget's affects the nipple and/or the areola, while eczema tends to spare the nipple. Eczema also tends to affect other parts of the body, or both breasts, while Paget's generally only affects one side. If untreated it can lead to an ulcer of the nipple. Symptoms of Paget's disease may include:

• Skin changes around or on the nipple such as crusting, flaking, and rash.
• The skin of the nipple or areola may be itchy, burning, or tingling.
• Nipple discharge.
• A lump behind the nipple.
• Nipple changes such as flattening or a nipple ulcer.

# gynecomastia

This condition is where the breast tissue grows larger than usual in men. Let's look at why this might happen.

Gynecomastia is the enlargement of the glandular breast tissues in males. The amount of breast tissue can be small—for example, around the nipples and areolae—or larger in size, and can affect one or both breasts. Although common, it can cause distress and affect self-esteem and mental health. It can also be tender and uncomfortable.

The breast tissue grows in response to estrogen. Males produce both the hormones estrogen and testosterone (as do females), but generally males produce significantly higher levels of testosterone than estrogen, meaning that the breast tissue does not grow. If this balance between estrogen and testosterone changes then gynecomastia can occur. This condition is common during puberty, affecting about half of male adolescents, as the levels of hormones change, but tends to disappear on its own within a few years.

As you get older, testosterone levels fall, meaning that gynecomastia is also more common in the elderly. Gynecomastia is not due to breast cancer.

## Possible causes

Having obesity can lead to gynecomastia, as fat cells are involved in the production of estrogen, meaning that the higher levels of estrogen may stimulate the growth of the breast tissue. Gynecomastia may also be a side-effect of medications, alcohol, and recreational drugs, or be related to other conditions such as low testosterone levels, hypothyroidism (low levels of thyroid hormone), and liver disease.

## Treatment

Treatment is typically not needed if the gynecomastia occurs during puberty, as it tends to resolve on its own within 6 months to 2 years. If it is related to a medication or recreational drug, stopping taking the medication may help. If the gynecomastia is related to an underlying disease process, then treating the disease may resolve the gynecomastia within a few months. Antihormonal medications may also be effective in treating gynecomastia, although they are not FDA approved for such.

Surgery can be considered in severe or persistent cases, or if the gynecomastia causes significant distress or discomfort.

# referral for further testing

You have been referred for additional testing. Let's look at what they are looking for and the most likely diagnoses.

If you find a change in your breast/s, talk to your doctor, who will ask questions and then might offer an examination. If you are under 30 years old, and the symptom (such as a breast lump, rash, or thickening) has appeared just before your period, your doctor may ask you to come back for another examination after your period. You might be referred for a mammogram. This doesn't mean that you have cancer; there are many things that can be going on, such as a benign breast cyst. It is natural to feel anxious about these kinds of tests, but it is important to rule out issues and get help if there are any.

## Referral to a breast center

You may be referred for additional imaging or tests by your doctor or radiologist if your mammogram demonstrates any areas of concern, or is inconclusive. These follow up appointments may be much longer than your initial screening, so bring a book or something to occupy yourself with as you wait. Typically, your appointment may start by reviewing your medical history, and possibly a breast exam. Depending on your imaging findings, a biopsy (or tissue sampling) may be recommended. If a biopsy is taken, that tissue will be examined by a pathologist. That report will usually be released to you through the breast imaging

center, or from the doctor who ordered your initial mammogram. Next we will look at some of the follow up imaging and procedures that you may need.

## DIAGNOSTIC MAMMOGRAM

Having a diagnostic mammogram is the same process as a screening mammogram, but this time additional images are taken of a specific area to get a better view.

## ULTRASOUND SCAN

Sound waves are used to create an image of the breast or breasts. You will be asked to remove the clothes on your top half and lie down with your arm above your head. Gel (often cold!) is applied to the breast and a probe rubbed over the breast to obtain the images. An ultrasound scan is not painful.

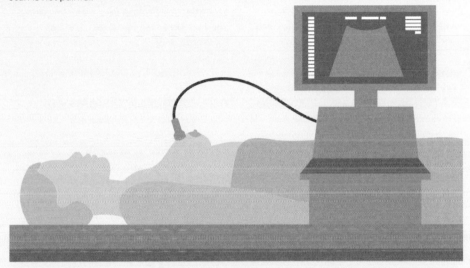

### • DON'T PANIC •

About 10 percent of women who have a screening mammogram in the US will be referred for further imaging. It is important to remember that most people who are referred don't have breast cancer. Being recalled after routine breast screening is more common after a first mammogram as there are no previous mammograms to compare the results with.

( . ) ( . )

## CORE BIOPSY

This is when a tiny sample of the breast tissue is taken using a hollow needle. Depending on where the sample will be taken, you may be asked to lie on your front or back. Local anesthetic will be injected into the skin to numb it. The local anesthetic feels like a bee sting or burning sensation before the area goes numb. Sometimes a small incision is made, then a needle put inside to remove a sample which is then analyzed by a pathologist. You may have more than one biopsy taken from the breast. A mammogram or ultrasound may be used at the same time to guide where to take the biopsy from. Typically, a small marker or clip is placed at the biopsy site to trace that area for any future changes. The biopsy site is then covered with a dressing, which needs to stay in place for a few days. After the local anesthetic wears off there may be some pain in the area, and you may notice a bruise. You can use over-the-counter pain relievers for any pain.

## FINE NEEDLE ASPIRATION (FNA)

Fine needle aspiration is used less commonly than core biopsies. An FNA uses a small caliber needle and syringe to take a small sample of breast cells to be examined. You may or may not be offered local anesthetic before an FNA, as the needle is so small the discomfort of the local anesthetic may be worse than that of the FNA itself. The FNA generally takes less than one minute per sample. A self-adhesive bandage is applied after the needle is taken out and can be removed later that day.

## SURGICAL BIOPSY

More rarely, surgery is needed to remove a sample for testing—for example, if a previous core biopsy result wasn't clear. In this case, you would be scheduled for outpatient surgery. Generally the whole lump would be removed as well as a small margin of healthy tissue.

( . )( . )

# benign breast changes

What might you discover at the breast center? All the conditions listed here are benign, meaning that they are non-cancerous.

### Fibrocystic breasts

Fibrocystic change is most common between the ages of 30 and 50, and involves small cysts (fluid-filled sacs) as well as areas of fibrous thickening. Symptoms include a general lumpiness of the breast, with lots of little lumps—often described as feeling like frozen peas—and this generally affects both breasts. It may also be associated with pain and tenderness. Fibrocystic change seems to be related to the menstrual cycle, in that symptoms are worse in the week or so before your period and then improve as your period starts.

### Fibroadenomas

Fibroadenomas are the most common type of benign breast lump in women between the ages of 14 and 35, and are an overgrowth of the glandular and connective tissues of the breast. These are benign and not cancer. A fibroadenoma generally feels round, firm, and rubbery and are sometimes called the "breast mouse" as they often move slightly under the skin on examination. They don't tend to be painful. They may resolve on their own, or, once diagnosed, can be removed but are often left because they are benign. If the lump continues to grow, or is large and affects the shape of the breast, then surgical removal may be advised.

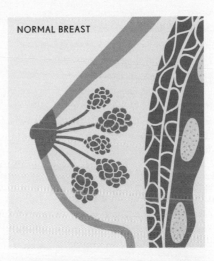

NORMAL BREAST

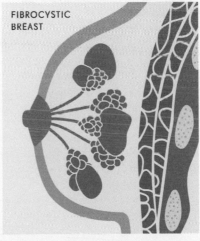

FIBROCYSTIC BREAST

Fibroadenomas often shrink or vanish after menopause due to the hormone changes involved (see page 116).

## Breast cysts

A cyst is a fluid-filled lump which feels smooth and firm. They can occur at any age but are more common around perimenopause and menopause. They can increase in size or first appear in the weeks leading up to your period and shrink or disappear afterward. They can be left once diagnosed, as they are benign, or the cysts can be drained (though they may refill).

## Lipoma

Benign lumps of fat overgrowth called lipomas can occur anywhere on the body, including the breasts, and can be left or removed. Lipomas often feel soft, like a lump which can be squashed, and can range in size. They can't be diagnosed without testing, so any breast lumps you find will need to be checked.

## Fat necrosis

The breasts contain fatty tissue, and if there is a trauma to this tissue, it can cause a lump known as fat necrosis. The skin around it might look red or bruised. It is more likely to occur with larger breasts and can occur in both men and women. Fat necrosis lumps can be left in situ, or removed if they are affecting the appearance of the breast or causing symptoms or concern.

## Sclerosing adenosis

A benign thickening or hardening, similar to scar tissue, within the breast. Sclerosing adenosis is benign, and does not change to malignant, so can be removed or left, once diagnosed.

## Intraductal papilloma

This is a lesion, which looks like a wart, generally found behind the areola. It can appear as a lump, and can also lead to a clear or blood-stained discharge from the nipple. These are removed to exclude cancer.

## Mammary duct ectasia

This is a benign condition which tends to affect women, but can occur rarely in men, where the milk ducts widen and thicken. It presents with a lump; nipple changes, such as new nipple inversion; nipple discharge; or with pain. If recurrent or persistent, surgical removal can be offered.

## Lumps related to pregnancy/ breastfeeding

These include blocked milk ducts, breast abscesses, galactoceles, and more. Although lumps related to breastfeeding are more common than cancer, breast cancer can occur during pregnancy or feeding so please get any changes checked. For more information, see chapter 5.

## Infections

Breast infections (mastitis) and abscesses most commonly occur during breastfeeding but can occur at other times. They are more likely if you have diabetes or if you are a smoker. They are usually treated with antibiotics, and abscesses may need surgical treatment. Having a nipple piercing can increase the risk of an infection or abscess. If you have symptoms of mastitis, without recently being pregnant or breastfeeding, or it does not improve within a week or so with antibiotics, you may be referred to a breast center; inflammatory changes in the breast may be related to inflammatory breast cancer.

## Mondor's disease

This is a rare, but benign, condition due to inflammation of a vein in the breast (thrombophlebitis). It can affect both women and men and may be related to exercise, breast injury, or breast surgery. Mondor's disease presents as a "cord" under the skin, which may be red and painful. It can then progress to looking and feeling like a tough, fibrous cord of tissue, which may look tethered. It resolves on its own; you can use over-the-counter pain relief until the pain improves (generally a few weeks), but the cord may take longer to disappear.

## Atypical hyperplasia

This is a benign overgrowth of cells which can occur in the ducts or lobes of the breast. Atypical hyperplasia is diagnosed on biopsy, and although it is benign, the cells look more unusual when compared to normal. Generally these are removed, as they are associated with an increased risk of breast cancer. You may be offered follow-up checks.

## Phyllodes tumor

This is a rare type of tumor and can be both benign or malignant, or be on the borderline of the two. Therefore, it is removed surgically. For information on breast surgery, see chapter 8.

> ### • LOBULAR NEOPLASIA •
>
> Lobular neoplasia or lobular carcinoma in situ (LCIS) is a condition which gives an increased risk of breast cancer. Depending on what exactly is seen, LCIS may be left or removed, as it does not seem to turn into breast cancer if left untreated. However, the increased risk means that you may be offered more regular screening or other treatments.

# 08

# Breast cancer

# risk factors

There is no definitive way to prevent breast cancer, but it's important to understand the risk factors linked with it, many of which are beyond your control.

Breast cancer is due to changes or damage to your DNA, which alters how your cells grow, divide, and replicate. These changes cause abnormal cells to multiply out of control. Some DNA changes are inherited, there may be damage from environmental or lifestyle factors, and in some cases the DNA damage occurs as the cells divide.

---

### • BREAST-CANCER SURVIVAL STATISTICS •

Breast cancer is the most common cancer in the UK and US, with 1 in 8 women in the US being affected at some point in their lifetime. With earlier diagnoses and better treatment pathways, thankfully this means that there are also many, many breast-cancer survivors, with around 80 percent of people with breast cancer surviving for more than ten years.

Currently it is estimated there are 3.8 million breast-cancer survivors in the US and around 600,000 in the UK, and it is predicted that these numbers will rise significantly.

---

Risk factors for breast cancer can be divided into non-modifiable factors (things you cannot change) and modifiable factors (things you may have some control over). But even if you have a non-modifiable risk factor, or lots of risk factors, it does not mean that you will definitely develop breast cancer. No matter the cause of the breast cancer it is important not to blame yourself, your genes, or any other factor; no one asks for cancer to happen. Whether or not you have risk factors and irrespective of your age, it is vital that everyone regularly examines their breasts or chest area. For self-check routines, see pages 39–43; for what to look out for, see page 40.

## Non-modifiable risk factors

Let's look first at the factors that you cannot control. All these factors increase your risk of developing breast cancer in your lifetime.

### BEING GENETICALLY FEMALE

Being born genetically female is the main risk factor for developing breast cancer. Breast cancer is more common in women, and around 55,000 cases of breast cancer are diagnosed in women each year in the UK, compared to around 350 men.

## AGE

The risk of breast cancer increases as you get older, with most breast cancers found in women over the age of 50. This is why many countries have breast-screening starting around this age (see page 119). In the UK, 280 women aged between 50 and 54 will be diagnosed with breast cancer per 100,000 women each year; compared to 412 new cases per 100,000 women aged 65-69; and 31 new cases per 100,000 women between the ages of 31 and 34 each year. About one quarter of breast cancer cases (24 percent) are in people over the age of 75.

## FAMILY HISTORY

The majority of people who have breast cancer do not have a family history of breast cancer. But having a close family history under a certain age increases the risk. Having a first-degree relative (parents, siblings, or children) under the age of 40 with breast cancer is linked with an almost double increase in risk. This risk is higher the more siblings are affected.

Around 5-10 percent of breast-cancer cases are hereditary, with genetic mutations passed on from one, or both, parents (see page 154). Genes which impact the risk of breast cancer can be passed down through your father's, as well as your mother's, side of the family.

If a woman has the BRCA1 or 2 gene (see page 156) they have approximately a 7 in 10 chance of developing breast cancer. There are also several other types of abnormal genes that increase risk of breast cancer. If you have a family history of breast cancer, you might be offered a test for BRCA and other genes (see page 155).

## PREVIOUS DIAGNOSIS

If you have had breast cancer before, there is a higher risk of developing a new cancer in the same or the other breast. This is a new cancer, not recurrence of a previous one. You may be offered more regular screening tests in order to detect a cancer as early as possible if it should occur. LCIS (see page 141) leads to around a 10-fold increased risk of developing invasive breast cancer in either breast, not just the breast where the LCIS was found.

## DENSE BREAST TISSUE

Breasts which contain more glandular tissue are denser than those which have more fatty tissue. Dense breasts increase the risk of breast cancer (there are more breast cells which can potentially turn into cancer), and makes it more difficult for mammograms to be interpreted (see page 122).

## HISTORY OF SOME BENIGN BREAST CONDITIONS

Certain non-cancerous breast conditions increase the risk of breast cancer, such as atypical ductal or lobular hyperplasia, where the ducts or lobules of the breast grow beyond the usual range. Lobular carcinoma in situ (LCIS) is a condition where there are cells which look like cancer—but aren't—growing outside of the lobule. Having LCIS does increase your risk of developing breast cancer. LCIS may have been picked up on a previous biopsy.

## EARLY PUBERTY AND LATE MENOPAUSE

Starting your periods early and/or going through menopause late, is thought to be related to a higher risk of developing breast cancer. This is due to having more exposure to the hormones of the menstrual cycle. The more times you ovulate in your life the higher your risk of developing breast cancer—for example, if you start your periods before 12 and have your menopause over the age of 55.

## HEIGHT

Taller women have a higher risk of breast cancer than short women; the reason for this is not yet understood.

## ETHNICITY

In the US, the incidence of breast cancer is higher in Black women than in white women. The data shows that Black women diagnosed with breast cancer are more likely to die from the disease. Black women are more likely to develop a triple-negative breast cancer. This is breast cancer which is estrogen- and progesterone-receptor negative as well as HER 2 negative (see page 164). This means that the cancer is harder to treat, may be more aggressive, and be more likely to recur.

The reasons for this are not fully understood and there are multiple factors involved including racial bias and disparities in access to health care. Black women are also more likely to develop a more advanced stage breast cancer. Compare the statistic that 22-25 percent of

Black women are diagnosed at stage 3 or 4 in the UK, as opposed to 13 percent of white women. One factor for this might be that Black women are less likely to be offered genetic counseling or testing: a study from Florida showed that Black women and Hispanic women were less likely to have genetic testing discussed with them by their doctors. Other reasons for lower testing rates may include lack of access to relevant specialists, language barriers, a lack of awareness, and economic factors such as health insurance. A further study, involving nearly 400 Black women who had breast cancer under the age of 50, showed that 12.4 percent had a BRCA1/2 mutation. However, over 40 percent of these women had not had a family history of breast or ovarian cancer. The disparities in health care for Black women with breast cancer need to be recognized and researched, in order for everyone to have the best treatment and outcomes possible.

## RADIATION THERAPY TO THE CHEST

If you have had radiation therapy at a young age—for example, to treat another cancer such as lymphoma—then your breast-cancer risk increases. The risk seems to be dependent on the age at which the radiation was given, with the greatest risk being when it was given under the age of 20.

## EXPOSURE TO DIETHYLSTILBESTROL

Diethylstilbestrol (DES) was given to some women between the 1940s and 1970s as it was thought to decrease the risk of miscarriage. If you took DES yourself, or had a mother who took it during her pregnancy with you, you might have an increased risk of developing breast cancer. There is around a 30 percent increased risk when compared to people not exposed.

## DIABETES

Women who have type 2 diabetes seem to be at a slightly increased risk of developing breast cancer, but we don't know why. Obesity and being overweight are risk factors for both type 2 diabetes and breast cancer, which may help explain the link.

## Modifiable risk factors

Let's now look at the modifiable or lifestyle risk factors that might increase or reduce your chances of developing breast cancer. Know that getting cancer is *never* your fault. You might do everything "right" (whatever that means) and still get cancer.

## WEIGHT

The relationship between breast cancer and weight is complicated. It is known that fat cells lead to inflammation, which can increase the risk of cancer. However, being obese or overweight before menopause is linked to a lower risk of breast cancer. After menopause (when you haven't had a period for a year, see page 116), the relationship between obesity and breast cancer seems to be a different story. Before menopause, the majority of the estrogen

made by the body comes from the ovaries, though some is made by adipose (fatty) tissue. After menopause, the ovaries stop working and no longer produce estrogen, so most of the estrogen comes from fatty tissue. Being obese or having excess weight means there is more fatty tissue, which produces more estrogen and this seems to be linked to an increased risk of breast cancer. Excess weight can also lead to higher levels of the hormone insulin, which is linked to a higher risk of breast cancer. Weight also seems to have an effect on whether or not the cancer is hormone-receptive positive.

### • BMI AND CANCER •

There is a lot of debate about BMI (body mass index) and how useful it is as a measure of health. However, it is currently used in medical research, and does give us some insight to the risk factor of obesity. For every 100 women of a healthy weight (a BMI of 20-25) over the age of 50, around 9 develop breast cancer. For every 100 women in the same age range, who have obesity (a BMI of over 30), around 11 or 12 will develop breast cancer, so an extra 2-3 cases per 100 women.

## EXERCISE

An active lifestyle with regular physical activity is associated with a lower risk of breast cancer. Some studies suggest that regular exercise can reduce your breast-cancer risk by 20-30 percent. Aim for 150 minutes of moderate-intensity exercise per week—for example, brisk walking, at a level of exertion where you can talk but don't have quite enough breath to sing. Or if you prefer high-intensity exercise such as HIIT or running, 75 minutes per week is the goal. Ideally, try to include weight-bearing and resistance-based exercise.

It doesn't have to be a specific exercise, or one of the ones mentioned here, and you don't have to be good at it; just find activities you enjoy and keep doing them. Whether it's swimming, dance, or lifting weights, just choose whatever you enjoy enough to do!

## DIET

Some studies have suggested that eating a diet rich in vegetables and fruit and low in processed and red meat may decrease the risk of breast cancer, but not all studies have supported this. More research is being carried out into diet and breast-cancer risk. The European Prospective Investigation into Cancer involves 10 countries and approximately 520,000 people, and is looking into lifestyle factors such as diet and cancer. A healthy balanced diet is of benefit to your general health and your gut health, and can lower the risk of other diseases.

## ALCOHOL

The risk related to alcohol and breast cancer increases with the amount of alcohol consumed. How alcohol increases the risk of breast cancer is not fully understood, but it may involve increasing the levels of estrogen in the body. Drinking just one unit of alcohol per day increases the risk of breast cancer by 7-10 percent, and drinking two or three units per day by about 20 percent, compared to women who do not drink alcohol.

Ensuring you stay within recommended limits of alcohol decreases your risk in comparison to people drinking more, but not as much as if you stopped drinking alcohol altogether.

### • HOW MUCH IS TOO MUCH? •

There is no known "safe" amount of alcohol to consume. If you do choose to drink alcohol, you should do so in moderation—1 drink or fewer per day for women, or 2 drinks or fewer per day for men. The amount considered 1 drink varies depending on the type of beverage:

• **Wine** (5 oz, ABV 12%*, 1.5 units)
• **Spirits e.g. gin, vodka, rum** (1.5 oz, ABV 40%*, 1 unit)
• **Beer or cider** (12 oz, ABV 5.2%, 3 units)

* ABV = alcohol by volume

## ALCOHOL AND BREAST-CANCER RISK

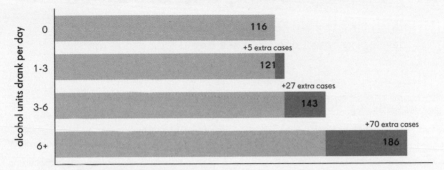

women diagnosed with breast cancer in their lifetime (per 1,000) in the UK

### SMOKING

There is some evidence that smoking increases your risk of developing breast cancer, and if you do develop breast cancer, increases your risk of recurrence. This is especially so if you started smoking during adolescence. The risk of breast cancer associated with smoking is also greater for women with a family history of the disease.

The benefits of stopping smoking are not confined to decreasing your breast-cancer risk, but include decreasing your risk of other cancers as well as decreasing your risk of cardiovascular and lung disease. Smoking is known to damage your DNA (including the parts of your DNA which protect against developing cancer), and the chemicals in cigarettes then impact the cells' ability to repair this damage.

### PREGNANCY

The relationship between breast cancer and pregnancy is complicated. Having pregnancies,

in particular at a young age (first pregnancy under 30), reduces the risk of getting breast cancer. If you have a pregnancy over the age of 30, or no pregnancies, it increases the risk.

The risk of breast cancer is higher for about ten years after being pregnant, and then recedes. For many people, whether or not you get pregnant is not a "lifestyle" choice.

You might have heard that miscarriage and termination of pregnancy may increase the risk of breast cancer, perhaps related to changes in hormone levels. Studies have shown this is incorrect; neither miscarriage nor abortions increase the risk of breast cancer.

### BREASTFEEDING

Choosing to breastfeed may reduce your risk of developing breast cancer. Breastfeeding for more than a year is associated with a slightly lower breast-cancer risk. This may be related to having fewer menstrual cycles when feeding. The ability to feed or not, however, is not always a choice.

# HRT, contraception, and breast cancer

Does taking HRT or using hormonal contraception increase your risk of developing breast cancer? Let's take a look at the facts ...

## Hormone replacement therapy

There has been a lot of high-profile debate in recent years about whether taking hormone replacement therapy (HRT) increases your risk of developing breast cancer. It depends on the type of HRT you take (it appears that it is the progesterone in HRT which is linked to breast cancer), your age, health, and other factors.

Age is important because age is a significant risk factor for breast cancer (after being female), so women in their 50s on HRT have a lower risk than older women.

### WITHOUT A WOMB

Women without a womb can use estrogen-only HRT, which has little or no impact on breast-cancer risk. Studies differ as to whether estrogen-only HRT protects against breast cancer, has a negligible effect, or a less protective effect than combined HRT.

### WITH A WOMB

Women with a womb need to be given both estrogen and progesterone in order to protect against womb cancer. There is no increased risk of breast cancer when HRT is taken for a year or less, and risk is related to the length of treatment.

Combined HRT has a higher risk of breast cancer than estrogen-only HRT. Even within combined HRT, the type of progesterone used affects the risk of breast cancer. Products containing the synthetic hormone norethisterone have the highest risk. Combined HRT with regulated, body-identical progesterone in the form of micronized progesterone (trade name Utrogestan) is not associated with an increased risk for the first five years of use, and then is thought to be linked with a smaller risk than with other kinds of progesterone. This risk also declines within approximately five years of stopping HRT.

There is no increased risk of breast cancer associated with the use of vaginal estrogen for genitourinary syndrome of menopause, which gives symptoms such as dry vagina, painful sex, and recurrent urinary-tract infections. Topical vaginal estrogen isn't the same as systemic

### • ALTERNATIVES TO HRT •

If you cannot have HRT (or choose not to) there are other non-hormonal prescription treatments and lifestyle options for menopausal symptoms. Please discuss with your doctor.

(whole-body) HRT and doesn't have the same risks. It isn't able to treat the whole-body symptoms you might experience around menopause, but can be life-changing for many women.

Really importantly, HRT does not increase your risk of dying from breast cancer. The small increased risk related to combined HRT is associated with "receptive positive cancer," a form which is generally easier to treat. This is not to belittle the impact of any kind of breast-cancer diagnosis, but it is an important piece of information to share in order for you to make your own informed decisions.

### PREMATURE MENOPAUSE AND HRT

If you have premature menopause and take HRT there are no increased risks until after the age of 50. This is because you are simply replacing the hormones that would otherwise be present until the average age of menopause.

### IF YOU ARE AT HIGH RISK OF BREAST CANCER

Whether or not you take HRT is an individual decision. Factors for consideration include the severity of your symptoms and the impact they are having on your life. It is important to assess your risk, as breast cancer in families is common (see page 145). If you are considered high risk, then HRT is generally (but not always) avoided. HRT is generally not recommended after breast cancer. Non-hormonal prescription options can be given, as well as lifestyle advice. In exceptional circumstances, however, it may still be issued.

## Hormonal contraception

The combined oral contraceptive pill is associated with a slightly increased risk of breast cancer, but only while you are taking it. The risk returns to the background population level within a decade of stopping. This small risk needs to be balanced against the benefit of contraception protecting against an unwanted pregnancy. The combined oral contraceptive pill also protects against womb and ovarian cancer.

Between ages 30 and 39, about 40 people per 10,000 will develop breast cancer. In the same age range of women who have taken the oral contraceptive pill for most of this decade, the number who will develop breast cancer increased to about 54 per 10,000 women. So, the combined oral contraceptive pill leads to an extra 14 cases per 10,000 women.

The number of times you ovulate in your lifetime seems to affect your risk of breast cancer—the more times you ovulate, the higher the risk can be. The combined oral contraceptive pill stops you ovulating, but the potential increase to breast-cancer risk is thought to be due to the fact that the hormones estrogen and progesterone are given. It is complex.

Other forms of hormonal contraception are progesterone-only such as the mini pill, implants, injections, and coils. It is not yet known if progesterone-only forms of contraception affect breast-cancer risk, and more research is required.

(.)(.)

# breast-cancer myths

### "WEARING A BRA CAUSES BREAST CANCER"

A suggested theory is that bras obstruct the flow of lymph in the body and can cause breast cancer, especially if wearing an underwire bra. There is no evidence to support this, nor is there a scientific reason why this would occur. Put simply, wearing a bra doesn't increase your breast cancer risk.

———

### "ARTIFICIAL SWEETENERS CAUSE BREAST CANCER"

There is no evidence to suggest that consuming artificial sweeteners, such as aspartame, causes breast cancer. Products containing artificial sweeteners are not considered harmful within daily limits. With regards to aspartame, one of the most common sweeteners, the recommended daily limit is 50 mg/2 lb of body weight. For example, a person weighing 150 lb, would have a recommended daily limit of 3,500 mg. To put that in context, the average amount of aspartame in a can of diet soda is about 200 mg.

———

### "ANTIPERSPIRANTS AND DEODORANTS CAUSE BREAST CANCER"

You might have read about the idea that toxins in antiperspirants and deodorants can enter the skin, build up, and cause cancer. There is no evidence to support this claim. Neither antiperspirants nor deodorants increase the risk of breast cancer.

———

### "MOBILE PHONES CAUSE BREAST CANCER"

There is no evidence to suggest that mobile phones, even when carried in a breast pocket or bag close to the breast, increase the risk of breast cancer.

———

### "STRESS CAUSES CANCER"

It has been suggested that stress increases the risk of cancer, in particular breast cancer. This is difficult to assess as psychological stress can impact lifestyle; for example, it can lead to increased alcohol consumption and decreased physical activity, both of which are associated with an increased risk of breast cancer. Stress can impact the immune system, which may affect cancer risk. However, currently there is no evidence to suggest that people who are stressed are more likely to develop breast cancer. Stress impacts us all, so it is important to try and look after your mental health and well-being.

———

### "BREAST IMPLANTS CAUSE BREAST CANCER"

Breast implants, be they made of silicone or saline, are not linked to an increased risk of breast cancer. Breast implants are, very rarely, linked to breast-implant-associated anaplastic large cell lymphoma (BIA-ALCL), which is typically found in the scar tissue around the implant. This is a form of non-Hodgkin lymphoma, which is a form of blood cancer. This is extremely rare, with fewer than 10 cases diagnosed each year, compared with hundreds of thousands of people having breast implants.

———

### "EATING SOY INCREASES YOUR RISK OF BREAST CANCER"

It was previously thought that eating soy-based products may increase your risk of breast cancer. This is because they contain isoflavones which are chemically similar to the structure of estrogen. In 2021, a review showed that although they are similar, they don't bind to all types of estrogen receptors in the body, therefore soy can be eaten without increasing the risk of breast cancer. Other research has shown that soy products may actually decrease the risk of breast cancer developing.

———

# genetic mutations

The human body is complicated, with thousands of genes that make you, you. We know that some genes are linked with increased risk of breast cancer, let's look at what this means.

You may be concerned about genetic mutations because of your family history of the disease, or you may have been informed by a relative who has had cancer that they are a gene carrier. It is thought that between around 1 in 10 to 1 in 20 cases of breast cancer are caused by genetic mutations. However, it is important to know that having a gene which increases your risk does not mean that you will definitely develop cancer. Breast cancer is the most common cancer in the UK, meaning that lots of people diagnosed with it will have a family member who is affected; this doesn't necessarily mean you have a hereditary gene increasing your risk.

## Family history

It doesn't have to be your mother who carries the faulty gene; it is anyone you share genes with, so if your sister has a gene mutation, you could have it, too, because you both got your genes from your parents. You may be monitored more closely if you have any of the following:

- A first-degree relative (a parent, sibling, or child) who had breast cancer under the age of 40.

- A first-degree relative of any age who had breast cancer in both breasts.

**NORMAL CELLS**

**CANCER CELLS**

cancer cells do not stop dividing and growing

cell sizes and shapes vary

larger darker nucleus

abnormal number/ disorganized chromosomes

cell boundaries poorly defined

- A male first-degree relative who had breast cancer, irrespective of age.

- A combination of family history: for example, if you had one first- or second-degree relative with breast cancer, and also a first- or second-degree relative with ovarian cancer; or two first-degree or one first- and one second-degree relative with breast cancer; or more than three first- or second-degree relatives with breast cancer, irrespective of age.

## Should I get genetic testing?

If you have a family history, you may be referred for genetic counseling. Genetic counseling is the first step. At the appointment, your family history is taken and the pros and cons of genetic testing are discussed with you. One important question to consider is what you will do with the information of the result.

For some people, knowing whether or not you are a gene-carrier helps end the uncertainty which may have been contributing to your stress and anxiety. For others, having a positive result increases their anxiety and they may prefer not to know that they have a gene. Remember that being a carrier does not mean that you will definitely develop cancer.

Some genetic tests may give inconclusive results. For example, a mutation may be identified of unknown significance, meaning at present it is not known whether or not it has an effect, which may also increase anxiety.

One of the pros to getting gene testing is that you may feel more in control, in that you can try and take steps to reduce your risk such as having

more regular screening, making lifestyle changes, and potentially having preventative treatment. Other factors to consider include how the result may impact families – for example, if one sibling is positive but another does not want to have testing (which is their right).

Genetic testing itself is quite quick and simple. It can be a blood or saliva sample, or a cheek swab, though getting the results can take some time. Although taking the test itself is simple, the implications of the result are not, so it is important to have the counseling and take time to consider your options.

## If you have a positive result

If you are found to be a carrier for a gene which increases your risk of breast cancer, you now have valuable information, which many people do not have. You may be referred into a high-risk screening program (see page 157) where you will receive specialist guidance and support. If you have a positive result, you may also wish to focus on the lifestyle risk factors outlined on pages 147–149. For example, limiting alcohol and staying active can help decrease your risk of developing breast cancer.

Depending on the result and your personal history and choices, you may be offered hormonal treatment (or "chemoprevention"). Hormonal medications such as tamoxifen or aromatase are used to block the impact of hormones which may be involved in the development of breast cancer.

You may also be offered preventative surgery which involves removing the breast tissue, known as a mastectomy (see page 165).

This reduces the risk of developing breast cancer by up to 97 percent, so almost completely. Depending on the gene identified, you may also be advised to consider surgery to remove the ovaries (oophorectomy). If you have the BRCA1/2 genes, this reduces the risk of breast cancer by about half if the ovaries are removed before menopause, and also reduces the risk of ovarian cancer. Removing the ovaries after menopause does not impact the risk of developing breast cancer, as the ovaries are no longer producing estrogen, but still decreases the ovarian-cancer risk. Many BRCA-related cancers are estrogen-receptor negative, so aren't influenced by estrogen.

## Types of genetic mutation

### BRCA1 AND BRCA2

The most common genetic changes causing hereditary breast cancer are the BRCA1 and BRCA2 gene mutations. These genes are involved in making proteins which help repair damaged DNA. In the mutated versions the genes lead to issues with cell growth, which may lead to breast cancer. The BRCA gene mutations occur in approximately 1 in 300-400 people. They are more common in the Ashkenazi Jewish population, occurring in around 1 in 40 people. It is thought that a woman who has a BRCA1 or BRCA2 gene mutation has up to about a 7 in 10 chance of developing breast cancer by the age of 80, compared to a 1 in 8 lifetime risk for the general population. Having one of these genes also means that there is an increased risk that breast cancer will occur at a younger age, or affect both breasts.

### PALB2 GENE

This gene makes a protein that interacts with a further protein made by the BRCA2 gene. PALB2 genetic mutations lead to a higher risk of breast cancer but are less common than the BRCA mutations.

### TP53

This gene helps stop cells with damaged DNA from growing. Inheriting a TP53 genetic mutation increases the risk of breast cancer and some other cancers such as leukemia.

### ATM

The ATM gene normally helps repair damaged DNA. Inheriting one damaged ATM gene can increase the risk of developing breast cancer. Inheriting two copies of the mutation causes ataxia-telangiectasia, which affects the nervous system and immune system.

## PTEN GENE

Inherited mutations of the PTEN gene can cause Cowden syndrome, a rare condition which increases the risk of breast and other cancers such as gastrointestinal and thyroid cancers. The PTEN gene usually helps regulate cell growth; if there is a mutation cancers can develop.

## CHEK2 GENE

This gene helps with repairing damaged DNA. Inheriting this mutation increases the risk of developing breast cancer.

## STK11

Mutations lead to a rare condition called Peutz-Jeghers syndrome, which leads to polyps in the gastrointestinal and urinary systems, leading to an increased risk of various cancers such as bowel cancer, as well as increasing the risk of breast cancer.

## CDH1

The CDH1 mutation leads to an increased risk of breast cancer and gastric cancer.

# Screening for people at higher risk of breast cancer

Routine breast screening tends to start at around the age of 50, although programs vary slightly between countries (see page 119). If you are at increased risk of developing breast cancer then you may be offered screening at an earlier age. If you think that you may be at higher risk—for example, if you have a family history of breast cancer—then please discuss with your doctor.

Whether or not you are offered earlier or more regular screening will depend on whether you carry a known gene, your age, and your family history. In the UK, if your family history means that you are considered to have a moderate to high risk of breast cancer, you will be offered annual mammograms from the age of 40. Under the age of 40, mammograms are less useful due to the density of the breasts, so you are may be offered an annual MRI scan instead.

If you are known to have a genetic mutation such as BRCA1 or 2 (see page opposite), you may be offered an annual MRI scan. The starting age for the screening may depend on the particular gene involved and may start from 20 or 30.

In the US, if you are considered to be high risk you may be advised to have an annual breast MRI and mammogram every year from the age of 30. These may be done together or alternately, every six months.

# breast cancer in men

Breast cancer can occur whatever your sex or gender,
though it is far more common in women.

In the US, 1 in 833 men will be diagnosed with breast cancer, and in the UK around 1 in 870. Compare that to around 1 in 8 women in the US, and 1 in 7 in the UK. The health outcomes of breast cancer in males is poorer than that of females, which may be related to the fact that they are more likely to be diagnosed at a later stage. Put simply, you might not be so aware that you are at risk of breast cancer, and are less likely to do regular checks. Men are also not automatically offered routine screening (see page 119).

The signs and symptoms of breast cancer in males are similar to that of breast cancer in females (see page 40). The majority of male breast cancers are estrogen-receptor positive (see page 164). There seems to be a higher risk of developing another breast cancer on the same side, or a breast cancer on the other side, than in women.

If you notice any changes to your chest area, please see your doctor.

## Treatment and research

Less than 1 percent of all breast cancers diagnosed each year are in men, and breast cancer in men is less than 0.5 percent of all cancer diagnoses in men in the US. This likely accounts for why the majority of breast cancer research is performed in women. This is in stark contrast to research conducted for most other malignancies.

---

### • GYNECOMASTIA •

This is a benign enlargement of male breast tissue, and is common (see page 135). Symptoms which may increase the likelihood that this tissue enlargement is related to cancer include: if it only affects one side; that it feels irregular or hard; and/or if there are enlarged lymph nodes in the armpit. The bottom line is: if there are any changes to your chest or breasts, get checked out.

(.)(.)

# types of breast cancer

There is an ever-growing body of knowledge about the types of breast cancer, which helps target the best treatment for patients.

## Noninvasive breast cancer

A noninvasive cancer, also known as carcinoma in situ, is a cancer which is still in the site in which it formed and hasn't spread. It is treated by surgery, though radiation therapy may also be offered. As there has been no invasive spread it has a better prognosis than other forms of breast cancer.

The cancer remains in the duct or lobule of the breast. Ductal carcinoma in situ (DCIS) is more common than lobular carcinoma in situ (LCIS) (see page 141). DCIS accounts for approximately 1 in 5 breast cancer cases, and is considered to be a precursor to invasive cancer so is removed.

## Invasive breast cancer

An invasive cancer is one which has spread from the tissue in which it started into the surrounding tissue. It may have started in the duct or lobule of the breast, and has grown and invaded into the surrounding breast tissue. The most common form of breast cancer (around 8 in 10 cases) is invasive ductal cancer, which starts in the ducts.

Approximately 1 in 10 cases are invasive lobular cancers, which start in the lobules of the breast. Invasive cancers are also further categorized by whether or not there has been spread into the lymph nodes or outside the breast into other parts of the body.

## Paget's disease of the breast

Paget's disease is a rare form of breast cancer which affects the nipple. There is commonly an associated carcinoma in situ, or invasive breast cancer underneath the affected nipple. For more information, see page 134.

## Inflammatory breast cancer

This is an aggressive, but rarer form of breast cancer. It is more common in younger women and in Black women, and unfortunately is often more advanced by the time of diagnosis, meaning that it may have a less favorable outlook. For more information on signs and symptoms of inflammatory breast cancer, see page 134.

# signs and symptoms of breast cancer

Not all breast lumps are due to breast cancer, so don't panic; but if you do notice any changes in your breasts, get assessed by your doctor.

Checking your breasts regularly and being aware of what is your "normal" is key to good breast health. Take a look at pages 39–43 for how to check your breasts. There are many reasons why you might develop a lump or rash; check chapter 7 for more information on the benign causes of these symptoms, but always get changes checked. Here we look at the possible signs and symptoms of breast cancer.

## What to check for

• A new lump or thickening in the breast.

• Changes to the shape or size of the breast. This could be a swelling of the breast, or of part of the breast. It could also be a puckering or tethering where the skin looks like it is being pulled from underneath.

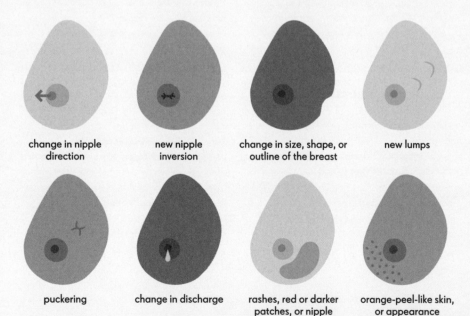

change in nipple direction

new nipple inversion

change in size, shape, or outline of the breast

new lumps

puckering

change in discharge

rashes, red or darker patches, or nipple crusting (ulceration)

orange-peel-like skin, or appearance of cellulite

- Nipple changes, such as the nipple changing direction, pointing inward, or nipple discharge.

- Skin changes such as redness or darkening of the skin. You might see dimpling of the skin, known as "peau d'orange," where the skin looks like orange peel or cellulite. Other skin changes include a new rash on the skin, or crusting on the breast or nipple.

- A mass or lump in the armpit.

- Pain in the breast or nipple (this is a rarer symptom of breast cancer).

## How is breast cancer diagnosed?

If you develop any of these symptoms, see your doctor for an exam. Your doctor will order any imaging you may need. If a diagnosis of breast cancer is made, then you will be referred to an oncologist. You may also need further imaging, genetic testing, and more.

### • WHAT ARE MY CHANCES? •

The first question that people often ask is: What are my chances? What is the survival rate? The answer depends on many factors, such as the stage of the cancer at the time of diagnosis, as well as your medical history and fitness level. There is not an exact answer. The best prognosis (the likely outcome) is for people who are diagnosed when the cancer is small and has not spread to other parts of the body.

Breast cancer survival rates depend on how far the cancer has spread. Localized breast cancers are associated with 99 percent relative survival rate, whereas breast cancer associated with distant spread has a 29 percent relative survival rate. This is why screening with mammography can be life-saving.

# staging of breast cancer

Doctors talk about cancer development in terms of stage, grade, and hormone-receptor status. What do these terms mean?

Depending on where you live, you may hear terms such as staging and grading of cancer, whether the cancer is primary, secondary, or metastatic, along with hormone receptor status.

There can be a lot of complicated medical language when you are diagnosed with cancer. Please always ask if you don't understand what your doctor is saying. The terms stage 4, secondary, advanced, or metastatic breast cancer all refer to the same thing. These terms indicate that the breast cancer has spread beyond the breast itself and local lymph nodes to other parts of the body. Less than 10 percent of breast cancers diagnosed are stage 4 at the time of diagnosis. In this situation you might hear that the cancer is "incurable," but this does not mean it is "untreatable." There are treatment options available and more people are living with cancer than ever.

## STAGES OF CANCER

Staging refers to the spread of the cancer. Most types of cancer are divided into four stages. You might also see the stages written in notes in Roman numerals.

Sometimes doctors use the letters A, B, or C to further subdivide the staging classification and define spread of disease. For example, stage 3B.

**STAGE 4 OR IV**

This cancer has spread to other areas of the body, such as the brain, bones, lungs, and liver. This invasive cancer is incurable, but not untreatable.

**STAGE 3 OR III**

This is a more advanced cancer which has spread beyond the immediate region of the tumor, and may include nearby lymph nodes or chest wall.

**STAGE 2 OR II**

The cancer is still contained in the breast or has extended only as far as the nearby lymph nodes

**STAGE 1 OR I**

The cancer is small and is contained locally.

| T | N | M |
|---|---|---|
| **T1**: tumor less than 2 cm across | **N0**: no cancer found in nodes | **M0**: no sign of cancer spread |
| **T2**: tumor 2–5 cm<br><br>**T3**: tumor over 5 cm<br><br>**T4**: tumor has spread into the skin or chest wall, or inflammatory breast cancer | **N1**: surgeon can feel swollen nodes, or cancer cells found in 1–3 lymph nodes in the armpit<br><br>**N2**: surgeon can feel nodes stuck together or attached to other structures, or cancer found in 4–9 lymph nodes in armpits<br><br>**N3**: surgeon can feel nodes near the collarbone or breastbone, or cancer is found in more than 10 armpit nodes | **M1**: cancer has spread to other parts of the body |

## TNM score

This is the staging classification most commonly used for breast cancers. The T refers to the primary (original) tumor size and how far, if at all, it had grown into neighboring tissues. This is graded out of 4. The higher the T number, the larger the cancer, or it has spread to the skin or chest wall. The N number (out of 3) refers to the number of lymph nodes, if any, affected. The M number (metastasized) refers to whether the cancer has spread to other parts of the body, and is shown as 0 or 1.

( . ) ( . )

## Grades

Your doctors might also refer to a grade, which categorizes the tumor on differing features of how the cells look under the microscope. Lower grade cancers are slow growing and less aggressive than high grade.

Some types of cancer have their own grading systems, but generally there are three grades:

**GRADE 1** the cancer cells look very similar to normal cells and are growing slowly.

**GRADE 2** the cells don't look like normal cells and are growing more quickly than normal.

**GRADE 3** the cancer cells look very abnormal and are growing quickly.

## Hormone-receptor status

Breast cancer is also evaluated for the presence of receptors on the cancer cells which allow proteins or hormones to attach to the cancer. The hormone-receptor status can affect which treatments are offered. For example, "triple negative" breast cancers do not contain receptors for any of the hormones or proteins below.

• **Hormone-receptors** Breast cancer may have receptors for the hormones estrogen and progesterone. Approximately 75 percent of breast cancers are estrogen-receptor positive. ER positive breast cancer has more treatment options available which act to block estrogen.

• **HER2** Human epidermal growth factor 2 is a protein which may be involved in the growth of cancer, but it is also found on normal breast cells. If the cancer has a particularly high level of HER2 receptors it is considered HER2 positive.

# breast-cancer treatment

Every treatment plan is unique and depends on the development of the cancer itself and your age, health, and personal choices.

Thankfully, there continues to be ongoing research and medical advancements in the treatment of breast cancer.

The treatment plan that you are offered for breast cancer will depend on the cancer, its stage, grade, and hormone-receptor status. Your general health and age, as well as shared decision making with your doctor, will also affect your treatment plan. Ask as many questions as you need, so the potential benefits and risks of any treatment options can be discussed and you can make an informed decision. If you think you might forget something, bring someone with you or take notes.

Depending on how advanced the cancer is, treatment options may have different goals. It may be aimed at curing the cancer, slowing the spread, or managing symptoms such as pain.

If cancer presents at a later stage of development, the focus of treatment might be to control the cancer and limit or slow down further spread. Or the aim of treatment might be to improve symptom control to reduce pain and enhance quality of life.

## Surgery

The type of surgery you may be offered will depend on the stage of the cancer, the size of the cancer compared to the size of your breast, the quality of the skin over your breasts, your general fitness for surgery, and whether or not you smoke. Quality of the skin depends on many factors including genetics, hormones, dryness, oiliness, and previous surgery in the area. Healthy skin heals better with less scarring. Smoking has a negative impact on wound healing so you are less likely to have a healthy healing response, which may impact the type of surgery offered.

### BREAST-CONSERVING SURGERY

A lumpectomy may be offered if the cancer is small relative to the size of the breast. The aim is to remove as much of the cancer as possible while conserving the remainder of the breast. This may also be referred to as a wide local excision. The cancer is removed along with some of the surrounding tissue to ensure that the surgical margins are free of disease. If it is appropriate, breast-conserving surgery is followed by radiation therapy, and the combination is thought to give an outcome equivalent to mastectomy. Recovery tends to be shorter than with a mastectomy.

### MASTECTOMY

This refers to the total removal of the breast. It is usually offered if the cancer is too large to be considered for breast-conserving surgery, or because it is in a place such as the middle of the breast which would make breast-conserving surgery difficult.

Breast reconstruction may be performed at the same time as the mastectomy or delayed (see page 165), and is considered part of breast-cancer treatment.

## SENTINEL LYMPH NODE BIOPSY

The lymph nodes which drain the breast can be biopsied to see if the cancer has spread to them. This is generally performed at the time of your breast surgery but may be scheduled separately.

## AXILLARY NODE DISSECTION

If the cancer has spread to the lymph nodes, most of the lymph nodes in the armpit will be removed surgically and checked to see how far the cancer has spread.

## BREAST RECONSTRUCTION OR PROSTHESIS

If you choose to have breast-reconstruction surgery this is considered part of your treatment. For more information, please see page 192.

If you have a mastectomy and choose not to have breast-reconstructive surgery, then you may choose to wear a breast prosthesis. For more information, see page 183.

### • HOW DOES YOUR TEAM DECIDE ON TREATMENT? •

Your doctors routinely communicate as a multidisciplinary team (MDT) which involves breast surgeons, medical oncologists (who use medications such as chemotherapy to treat cancer), radiation oncologists (who use radiation therapy), radiologists, specialist nurses, and more. The purpose of MDT communication is to utilize the expertise of all these different specialists in order to decide the optimum treatment plan options in your individual case.

One of the tools used in the UK may include a computer algorithm known as Predict Breast; this is based on the data of thousands of women and their tumors. It calculates the benefits of chemotherapy, hormone therapy, and immunotherapy to give a 5- or 10-year prognosis to help with decision-making.

As you start and continue your treatment, your response to the treatment—and any side-effects—will also inform decisions about your treatment.

## Radiation therapy

Radiation therapy aims to kill cancer cells by stopping them multiplying. It usually involves beams of high-energy radiation and is often used in addition to breast-cancer surgery to prevent the cancer recurring. You might hear radiation therapy called adjuvant therapy, if it used after surgery.

Usually radiation therapy is started four to six weeks after breast-conserving surgery and involves going to the hospital for daily treatments for a period of a few weeks.

Before you have radiation therapy you will be asked to have a planning CT scan of the area. Small, permanent dots (tattoos) are made on the area so that the correct area is exposed to the radiation.

If your breast cancer is on the left side, you may be asked to inhale and hold your breath briefly during the radiation therapy sessions as this moves the heart away slightly from the area having the treatment.

Other options for radiation therapy include having it administered internally, either during surgery, or using brachytherapy, whereby small tubes containing radioactive material are put into the area where the cancer was, for a short period of time, before being removed.

### SIDE-EFFECTS

The high-energy radiation destroys cancer cells but can also damage normal, healthy cells which may lead to side-effects. Side-effects will usually improve within a few weeks of stopping radiation therapy. These include:

- **Fatigue** You may experience fatigue during and for up to two months after radiation therapy, though fatigue can be a longer-term issue after treatment for breast cancer. Rest when you need to, but also try to do some physical activity as this can actually give you energy!

- **Skin irritation** The skin exposed during radiation therapy can get red, itchy, sore, and dry. There may be blistering or burning (known as radiation burn or radiation dermatitis) or it may become darker in color. In rare cases there may be permanent skin staining. Avoid being in the sun and wear high-protection sunscreen for at least a year after treatment.

- **Pain** There can be pain in the area of treatment, which may also swell.

- **Changes to the breast itself** Longer-term side-effects of radiation therapy include changes to the breast, which may feel firmer or become smaller. There can be damage to blood vessels in the skin, for example, you may notice "spider veins," small red marks on the skin.

---

**• RADIATION THERAPY CONCERNS •**

Radiation therapy isn't painful; you can't feel it. You might feel uncomfortable lying in the required position under the machine. Having radiation therapy doesn't mean that you yourself become radioactive, so it is safe to be with other people after treatments, including children and pregnant women.

---

• **Chest pain** Very rarely, radiation therapy can affect the heart or lungs if the left side was treated. It might affect the ribs. If you have chest pain or shortness of breath, seek medical advice.

## Chemotherapy

Chemotherapy is a form of medication which kills cancer cells, or slows or stops cancer cells multiplying. Chemotherapy may be given before surgery with the aim of shrinking the tumor so surgery can be more safely performed, which is known as neoadjuvant therapy. Or chemotherapy might be given after surgery, destroying any cancer cells too small to be visible to the naked eye, which may have spread from the area which has been removed. This is known as adjuvant therapy.

Chemotherapy can also be used in some circumstances if the cancer has spread from the breast to other parts of the body.

The type of chemotherapy offered will depend on the type of breast cancer and is generally given intravenously, though some forms of chemotherapy are taken orally. You may be prescribed a combination of different chemotherapy medications because they work in different ways.

### SIDE-EFFECTS

The side-effects that may occur will depend on the type of chemotherapy you are offered.

A common side-effect is hair thinning or hair loss—but this doesn't always occur. This can affect all the hair on your body, not just on your head. Wearing a cold cap to reduce the blood flow to the scalp and reduce the effect of the chemotherapy on the hair works in about half of people using it. Other side-effects include nausea; vomiting; indigestion; diarrhea; constipation; headaches; mouth ulcers; dry, itchy skin; muscle and joint aches and pains; and fatigue. Some people experience "Chemo brain," which is difficulty with memory and concentration.

Talk to your oncologist so they are aware of any side-effects you are experiencing and you can get the best treatment options for you. Generally, side-effects can be managed well with anti-nausea or anti-diarrhea medications.

Ensuring that you stay well-hydrated can help; and exercise, which includes walking, can help reduce the side-effects of chemotherapy.

Chemotherapy can affect the bone marrow, leading to a reduced number of white blood cells, which help fight infections. This means you will be more susceptible to infection, and you should contact your doctor urgently if you develop a fever. During chemotherapy patients also have a reduced number of red blood cells which can lead to anemia, and blood transfusions may be recommended.

## Immunotherapy

Immunotherapy is a form of treatment which stimulates the immune system to fight the cancer. There are different types available. Trastuzumab (also known as Herceptin) is offered to people with HER2 receptor-positive breast cancer. It is a monoclonal antibody medication which attaches to the surface of the cancer cells, stopping them from multiplying.

It is given alongside chemotherapy. New therapies include CDK4/6 inhibitors, such as Palbociclib, which are targeted therapies for secondary breast cancer. These work by interfering with a protein involved in how cells divide and multiply. It can be taken orally or through intravenous infusion.

# Hormone treatments

The receptor status of the breast cancer is assessed to see if the cancer responds to, and is affected by, the hormones progesterone and estrogen. If it is, medications can be used to lower hormone levels with the aim of preventing the cancer recurring. Although estrogen and progesterone are mainly made in the ovaries before menopause, estrogen is also made in other tissues, for example, in fatty tissue. Occasionally, hormone treatment is also given for non-hormone-responsive cancers. These medications can affect not only the breast but the rest of the body, causing menopausal-type symptoms.

## ESTROGEN BLOCKERS

The most commonly used hormone treatment is tamoxifen (a selective estrogen receptor modulator), which works by blocking cells from using estrogen. This is usually taken for five to ten years after breast cancer treatment for primary breast cancer which hasn't spread. Tamoxifen is often given if you have not been through menopause, but can be given at any point. If it is being used for secondary, or advanced breast cancer where the cancer has

spread to other parts of the body, it can be used for as long as it continues to be effective.

## AROMATASE INHIBITORS

This hormone treatment lowers your estrogen levels. Examples include letrozole, anastrozole and exemestane, used by women who have gone through menopause, meaning that their ovaries are no longer producing estrogen, though some is still made, generally from fatty tissue. Aromatase inhibitors stop these other tissues from producing estrogen. Some women are switched from tamoxifen to an aromatase inhibitor after they go through menopause.

## GONADOTROPHIN-RELEASING HORMONE ANALOGUES

GnRH analogues, such as goserelin (Zoladex) injections, can be used for those who have not yet gone through menopause. They work by stopping ovulation and therefore the production of estrogen and progesterone. It can be used alongside other treatments such as aromatase inhibitors.

## TREATMENT AFFECTING THE OVARIES

Treatment to remove the ovaries surgically, or stop them working using radiation therapy, is also considered a hormone treatment. The result of these treatments is that your body won't be able to make estrogen anymore.

# prehab—before treatment

Diagnosis and treatment can be a very stressful time. Let's look at what you can do during this stage to keep feeling as well as possible.

Prehabilitation (prehab) essentially means working to optimize your physical health before your treatment starts. This might mean stopping smoking if you are a smoker, limiting your alcohol intake, eating a healthy diet, and doing physical activity. All of this can help you cope well with the side-effects of treatments, and will aid your recovery.

## Exercise

It might feel like exercise is the last thing you want to do, while you have breast cancer, but there are many benefits. You might be concerned about whether exercise is safe for you. We all need to keep physically active, for the myriad health benefits that exercise provides, both physical and psychological. This is also true with regards to cancer and its treatments. If you are undergoing cancer treatment, or living with cancer, there are significant benefits to regular exercise, all contributing to your quality of life. These include improved mental functioning; reduced fatigue, anxiety, and depression; and better physical health with improved cardiovascular and musculoskeletal health. The American Society of Clinical Oncologists produced guidance in 2022 which specifically stated that doctors should recommend physical activity to people receiving treatment for non-metastatic breast cancer. So, if you already enjoy regular exercise, then do carry on if you feel able to. Maintaining your usual routine can help reduce feelings of stress and anxiety, and help with sleep. If you are concerned about the type of exercise you are doing, then do discuss with your doctor.

Some studies show that people who exercise regularly (on average 150 minutes per week of moderate intensity exercise) reduce the risk of recurrence and death from breast cancer by about 50 percent. If you are new to exercise, start with gentle exercise like walking and stretching classes. Physical benefits aside, exercise will help with sleep, release endorphins to make you feel good, and can help lift your mood.

# Regular exercise can reduce the risk of recurrence and death from breast cancer.

( . ) ( . )

# rehab—after treatment

Be gentle with yourself, you have just had surgery
and your body will need time to heal.

Rehabilitation (rehab) refers to recovery after treatment. Make sure that you are eating and drinking well and getting enough rest. Keep walking regularly as this helps you maintain your activity levels and also helps with other issues such as constipation, which may be a side-effect of pain relieving medication. There are lots of treatments available for side-effects so please don't suffer in silence; talk to your medical team.

Avoid lifting anything heavier than a gallon of milk on the side of surgery for the first six weeks after your operation. Avoid any strenuous exertion for six weeks before gradually returning to exercise. If you notice any activity is causing pain or increasing swelling in the area, then stop. For example, even gentle repetitive movements like knitting may cause pain and swelling. Depending on the extent of the surgery you will also be given advice about when you can start driving.

## Recovery exercises

After breast cancer surgery or radiation therapy you may be given exercises to help with recovery. There are many benefits to these including:

• Helping you regain the movement in your arms and shoulders.

• Reducing cording (see page 178).

• Reducing and preventing back pain after surgery.

• Aiding circulation and healing.

• Preventing long-term mobility and stiffness problems.

• Reducing the risk of lymphedema (see page 177).

• Improving flexibility and reducing stiffness ahead of radiation therapy so you can move your arm into the required position to avoid delaying treatment.

### WHEN TO START

Your surgeon or physical therapist will give you personalized information and guidance—do not start exercises until you have been given permission by your health care team. Depending on the type of surgery, the exercises may be started as early as the day following surgery. You then continue the exercises until mobility returns to normal, which may take a few months. If you develop any complications, you may be advised to take a break.

Start with the deep-breathing exercise below, which can be done from any point and as many times as you like. Generally you will be advised to start with the gentle exercises and progress to the more advanced ones, but the exact timing of this, and which exercises to do, will depend on your surgery and treatment. For example, you may have drains or sutures, you may have had reconstructive surgery, or your surgery may have involved the armpit. The exercises below are simply common examples. You will be told which are suitable for you, when to do them, and how many repetitions to do by your health care team. Take it slowly, and be gentle on yourself.

# GENTLE EXERCISES

## DEEP BREATHING

This is good for opening your lungs and helping you relax. You can do this lying down or sitting up—whatever is comfortable for you.

**01.** Breathe slowly and deeply in through your nose and let your chest and tummy expand as you do so.

**02.** Exhale slowly through your mouth. Continue for a few minutes.

## SHOULDER SHRUGS AND CIRCLES

This exercise is good for relaxing any tension and stiffness you are carrying in your neck and shoulders. Do these sitting or standing; whatever feels good for you.

**01.** Shrug your shoulders up to your ears and then relax down again. Keep your arms loose and relaxed.

**02.** As you lift your shoulders, try breathing in. As you relax them down, breathe out. Repeat as many times as directed.

**03.** This time, lift your shoulders up toward your ears and then circle them back down and round. Keep your arms loose and relaxed.

**04.** Repeat as advised, letting your shoulders relax a little more each time.

### • SAFETY TIPS •

For the first two weeks after surgery, do not lift your arms higher than your shoulder. The exercises should not be painful, but you may feel a stretching or pulling sensation. Do not try to "push through" pain or discomfort; instead gradually increase the stretch over weeks.

Remember, even brushing your hair can be a good exercise!

## ARM BENDS

These help you get some easy movement for your elbows, arms, and shoulders.

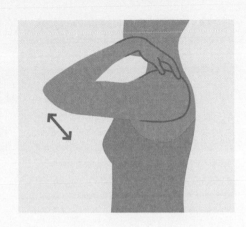

**01.** For a forward elbow bend, bring your hands up to rest on your shoulders. With your elbows bent, raise your arms forward, so they are at a right angle to your body. Do not raise your arms above your shoulder. Slowly lower your elbows again.

**02.** For a sideways elbow bend, start with your arms turned outward and bring your hands up to rest on your shoulders so that you bend your elbows out to the sides. Slowly lower your elbows down to your sides.

**03.** Repeat as many times as advised, relaxing your arms and elbows down each time.

## "FASTENING YOUR BRA" STRETCH

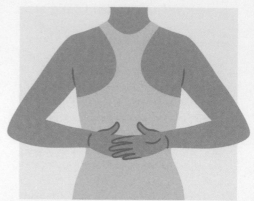

**01.** Hold your arms out to the sides with relaxed, bent elbows (like a scarecrow).

**02.** Reach behind your back as if you were fastening a bra or scratching your back. Then lower and relax. Repeat as advised.

If you find this tricky, do one arm at a time initially, and build up to doing both at the same time. Remember not to push yourself beyond what is comfortable.

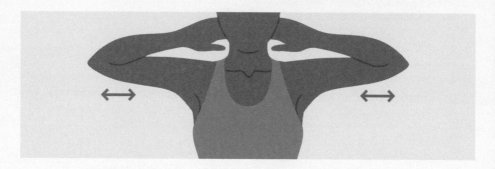

## SUNBATHER STRETCH

You can do this sitting or lying down. If lying down it will take time for your elbows to be able to open out toward the floor. Only do what is comfortable for you.

**01.** Lift your arms and place your hands on the back of your head with your elbows pointing forward. Then open your elbows up as if you were sunbathing.

**02.** Return your elbows to the front and repeat as directed.

**03.** When you feel comfortable doing this while sitting up, keep your elbows out to the side, and bend your body from side to side.

## ADVANCED EXERCISES

### WALL CLIMBS

**01.** Start by standing facing a wall with your feet slightly apart and put your hands on the wall at shoulder height. Gradually walk your fingers or slide your hands up the wall until you feel stretched (not painful), hold for a count of ten and walk back down before repeating, aiming to get higher each time.

**02.** Then repeat sideways. Turn so that the side you had the surgery on is facing the wall. With your elbow bent, put your hand on the wall and walk your fingers or slide your hand up the wall, straightening the elbow as you go. When you feel a stretch (not pain), stop and hold for a count of ten before sliding back down and repeating. In both of these exercises keep your head straight, looking forward, not twisting or turning.

**03.** This motion can also be done forward, sitting on a chair close to a table and walking your hand forward.

### ADVANCED SUNBATHER STRETCH

**01.** Lie down with your hands behind your head and elbows out to the sides like you are sunbathing.

**02.** Push your elbows down into the floor, hold for a count of ten, and relax.

### OVERHEAD STRETCH

You can do this sitting or lying down.

**01.** Relax your shoulders down and clasp your hands together in front of you.

**02.** Bring your arms together up and over your head to stretch. Hold for a count of ten before returning to your start position. Repeat as directed. You may find this easier to do holding a stick or ruler between your hands.

### SHOULDER SQUEEZES

You can do this sitting or standing.

**01.** Place your hands lightly on your shoulders.

**02.** Squeeze your shoulder blades together so your elbows move backward. Take care not to shrug your shoulders up at the same time. Repeat as advised.

### SIDE BENDS

You can do this sitting or standing.

**01.** Slowly lift your arms above your head and clasp your hands together.

**02.** Now slowly bend to each side in turn.

# wound care

You are back home and on the road to recovery. Here
is how to care for your wound from surgery.

After your operation, the incision will likely
be covered with a surgical dressing. You will
be given instructions on when you can shower
and bathe, and when to remove the dressing. It
may be that you have to return for sutures or
staples to be removed and dressings changed
(or you may still be in the hospital).

Once the dressings are removed and the skin
has healed, you will be given instruction for
further care if needed, which generally includes
to wash the area as normal. Rinse the area well
and pat the wound and surrounding skin dry;
don't rub the wound. Always use new dressings,
gauze, and bandages; don't try to reapply used
ones. Don't pick at or try to remove scabs.

## Surgical complications

All surgery has a risk of bleeding, blood
clots, and wound infections, as well as other
complications, though precautions will be
taken to try and decrease this risk.

There is also the risk of a seroma, which is
a collection of fluid in the tissues around the
site of the surgery, or hematoma, which is a
collection of blood. You may notice swelling
which doesn't go down initially, which can lead
to discomfort and restricted mobility. Seromas
tend to absorb on their own after a period of
time; however if they are very painful they can
be drained.

## Massage

Once your scar has healed it may be helpful to
massage it to ensure that the scar stays supple
and flexible and doesn't restrict movement. Use
a simple emollient and the flat of your fingers or
palm of your hand to rub the skin of the scar.
Gentle massage can also help treat any pain or
other sensations in the scar such as tingling;
numbness may improve but may also be
permanent. If you have scars that you cannot
reach, for example on your back, ask someone
else to massage them. Your surgeon may also
advise on treatments to reduce the appearance
of scars, such as micropore or silicone tapes.

### • RED-FLAG SIGNS •

If you develop signs of infection such as
worsening swelling, redness, heat, or pain,
please seek urgent medical attention.

# complications after treatment

There can be complications following breast-cancer
surgery. Let's look at how to reduce your risk.

## Lymphedema

Lymphedema is a buildup of lymphatic fluid,
leading to swelling of the arm and hand after
breast-cancer surgery. This is due to damage to
the lymphatic vessels or removal of the lymph
nodes during surgery. The risk of lymphedema
will depend on the extent of the surgery, with
axillary node dissection having two to three
times the risk of sentinel node biopsy. Radiation
therapy also increases your risk.

Lymphedema affects around 1 in 5 after
breast-cancer surgery. Although lymphedema
can develop rapidly, within days of surgery, it
can also take months or even years to occur,
so if you have concerns at any point, talk to
your doctor.

It is normal for there to be some swelling in
your chest and arm for the first four to six weeks
after surgery, so let's look at the key symptoms
of lymphedema.

### SYMPTOMS

Symptoms of lymphedema include swelling
in any part of the hand, arm, or chest, which
might feel tight and uncomfortable, stiff,
inflexible, or heavy.

Clothes or jewelery may feel tight or difficult
to get on or off.

## REDUCE YOUR RISK

• Try to move your arm as usual and keep
physically active. The exercises on pages
172-175 may help improve mobility.

• Eat a healthy, balanced diet, and avoid adding
too much salt to your food. Drink plenty of
water to stay well-hydrated.

• Try to avoid new potential skin infections as
this can worsen lymphedema: for example,
wear gloves when gardening, use insect
repellent to avoid getting insect bites, keep
any cuts or grazes clean, and keep skin
well-moisturized.

• Use sunscreen. Getting sunburned can
increase your risk of developing lymphedema.

• You can have your blood pressure and
blood tests taken from the affected arm, but
if possible, the other arm is used.

### TREATMENT

You may be referred to a lymphedema
specialist or physical therapy for management.
Unfortunately, there is not yet a cure for
lymphedema. You can, however, alleviate your

symptoms. Experimental treatments are being developed to try and move lymph nodes from other parts of the body into the armpit, or reconstruct the lymphatic vessels, so we hope treatment will improve in the future. Treatments for lymphedema include:

- **Manual lymphatic drainage (MLD)**, a form of massage which aims to help the fluid drain away from the arm.

- **Compression garments** such as gloves, sleeves, or bandages. Be careful taking them on and off to make sure that they don't roll and that there are no wrinkles, as this may lead to too much pressure being applied.

- **Kinesio-taping** is suitable for some people; taping aims to encourage the movement of lymphatic fluid.

## Cording

Cording, or axillary web syndrome, can occur after breast-cancer surgery. The cause is not fully understood but is thought to be due to tiny fibrosed (scarred) lymph channels. Tight webs or cords form from the armpit down the inner upper arm and can cause pain and discomfort.

### SYMPTOMS

Cording may feel painful, or you may feel that your arm movements are restricted, for example, you can't get your arms above your head.

### TREATMENT

Many of the exercises on pages 172–175 will help alleviate cording. You may notice a snapping sensation, or hear a pop as you do the exercises, which isn't painful—this is the cords releasing. You may then find that you can move your arm more normally. A physical therapist can also massage and help stretch the area to release the cords.

## Post-mastectomy pain syndrome

Post-mastectomy pain syndrome (PMPS) affects 20–60 percent of women who have a mastectomy, though it can also occur after breast-conserving surgery. It is most common after surgery which removes tissue in the upper outer section of the breast or the underarm region.

In PMPS there is chronic pain after the surgery, and although it may improve, the pain may not ever resolve completely. The exact reason for PMPS is not known but it may be related to nerve damage in the armpit or chest during surgery.

### SYMPTOMS

The pain itself may be achy, tingling, prickling, itching, numbness, or burning over the chest, armpit, shoulder, and upper-arm areas.

### TREATMENT

PMPS is treated with pain relieving medication or neuromodulators, as well as chronic-pain management therapies such as psychological therapy.

# recurrence and secondary cancer

Patients often worry, very understandably, about breast cancer returning after treatment. Here is what to look for.

## Local recurrence

It is important to continue to be aware of any changes in your breasts, even if you have had a mastectomy (for more information on breast examination see pages 39–43).

If the breast cancer comes back in the same breast it is called local recurrence. This is most likely in the first five years after diagnosis.

The risk of local recurrence in the first ten years is between 2 and 15 percent following breast-conserving surgery (a lumpectomy) and radiation therapy. Recurrence may be picked up during a follow-up appointment or scan or you may notice symptoms.

### SYMPTOMS

- A lump on the breast or on the scar from previous surgery. This may be pink, red, or darker in color.
- A new lump or area of thickening in the breast.
- A change in the shape, position, or direction of the nipple.
- A change in size or shape of the breast itself.
- Skin changes such as redness or a rash on the breast or nipple.
- Crusting on the nipple or breast.
- Swelling in the arm or hand of the affected, treated side.

If you notice any changes or have concerns, please see your oncologist so that you can be assessed.

### TREATMENT

If there is local recurrence, treatment options include surgery to remove the whole breast (mastectomy), if previously you had breast-conserving surgery (a lumpectomy); radiation therapy; chemotherapy; immunotherapy; and hormonal treatments.

## Secondary breast cancer

Secondary breast cancer is when the cancer, which originated in the breast, has spread (metastasized) to other parts of the body.

> ### • "SCANXIETY" AND ANXIETY •
>
> Anxiety about cancer returning often peaks toward the time of follow-up scans. Getting support and cognitive behavioral therapy can be useful.

Having one or more of the symptoms listed here does not mean that the cancer has spread (and many of these symptoms are nonspecific); rather it means that you should see your doctor to be assessed.

If breast cancer spreads, it tends to spread to the bones, lungs, brain, or liver. Treatment may depend on the site of the spread and can include radiation therapy, chemotherapy, immunotherapy, and hormone treatments. At this stage, treatment is no longer aimed to be curative but to delay spread and slow growth.

## GENERAL SYMPTOMS

• Fatigue.
• Unintentional weight loss (when you haven't been actively trying to lose weight, for example, you may notice that your clothes feel looser on you).
• Nausea and loss of appetite.

## SYMPTOMS OF BONE METASTASES

This is when cancer has spread to the bones. The bone marrow is involved in the production of blood cells and platelets, so symptoms may be related to the lower levels of blood cells and platelets, such as:

• Pain in your bones.
• Bone fractures.
• Fatigue and shortness of breath due to anemia.
• Bruising or bleeding due to low platelet levels.

## SYMPTOMS OF LUNG METASTASES

• Shortness of breath.
• Persistent cough.
• Chest pain.

## SYMPTOMS OF BRAIN METASTASES

• Headache.
• Nausea and vomiting.
• Loss of strength or sensation in part of the body.
• Loss of balance.
• Changes in vision.
• Confusion.
• Seizures.
• Changes in personality.

## SYMPTOMS OF LIVER METASTASES

• Abdominal pain.
• Hiccups.
• Jaundice.
• Abdominal swelling due to a buildup of fluid (ascites).

## SYMPTOMS OF SKIN METASTASES

Skin lumps or changes such as redness or swelling of the skin.

# breast-cancer survivorship

After your treatment is complete, your general health and mental well-being will also need time and support to recover.

Currently, there are around 3.8 million breast-cancer survivors in the US and 600,000 in the UK. With early diagnosis and better treatment pathways, it is predicted that these numbers will increase as more people continue to thrive despite living with breast cancer.

## Mental health

A cancer diagnosis can, of course, impact your mental health, and this may persist long after treatment. There are lots of treatments available, such as counseling, antidepressants, and anti-anxiety medications.

"Chemo brain" or "brain fog" are often used by patients to describe difficulties with memory and concentration that can occur when on chemotherapy and beyond. The cause after treatment is ended is unclear; it could be related to the breast cancer itself, chemotherapy, or menopausal symptoms.

Physical exercise, mindfulness, and yoga may be useful to support you during this time. There are also likely to be local and national support groups you can access (see page 197).

## Fatigue

Fatigue is extremely common after breast-cancer treatment, with around half of all patients experiencing fatigue within the first five years. It is unclear why this occurs and may be related to the impact of treatment, the mental-health implications of a cancer diagnosis, and more.

Fatigue is more likely to occur if you have had chemotherapy or have a mental-health condition such as depression or anxiety.

Treatments for fatigue include exercise, cognitive behavioral therapy, and peer-support groups. Although it sounds contradictory, physical activity actually helps fatigue, and also helps decrease recurrence of breast cancer.

## Lymphedema

This can develop quickly after breast-cancer surgery or occur later. See page 177 for more detail.

## Bone health

Breast-cancer medication which blocks estrogen can lead to an increased risk of osteoporosis, a condition in which the bones become thinner, fragile, and brittle, with an increased risk of fractures.

This is most associated with aromatase inhibitors; if you are on this medication it is likely that you will be offered a bone-density scan. Depending on the results you may be offered treatment. You may be advised to take a vitamin D and calcium supplement, eat a diet rich in calcium, and perform weight-bearing exercise to protect your bone health.

## Infertility

Chemotherapy, radiation therapy, or surgical removal of the ovaries can all lead to infertility.

## Menopause symptoms

Chemotherapy can affect the ovaries, causing a premature or early menopause. The ovaries may also be affected by radiation therapy or removed surgically. Hormone treatments, such as tamoxifen and aromatase inhibitors, can temporarily cause symptoms of menopause as they block the production or effect of estrogen.

Menopausal symptoms can have a significant impact on your quality of life and may be more severe or intense than in those going through a physiologic, or "natural," menopause. Here are just some of the symptoms:

• Hot flashes and sweats.
• Joint pains.
• Headaches.
• Dry, itchy skin.
• Fatigue and insomnia.
• Loss of libido.
• Depression, anxiety, and irritability.
• Difficulties with memory and concentration, and more.

Genitourinary symptoms related to hormone changes can also occur, such as genital dryness, soreness and itching, painful sex, and recurrent urinary-tract infections. These symptoms are most common following treatment with aromatase inhibitors such as anastrozole.

## TREATMENT

Hormone replacement therapy (HRT) is generally not given to breast-cancer survivors. In exceptional circumstances, after discussion of the risks and benefits, and after trials of other medications, sometimes it is sometimes considered. However, this refers only to whole-body, or systemic, HRT. Vaginal estrogen is an effective treatment for genitourinary symptoms of menopause, as it acts locally and is not absorbed into the rest of the body. In general, this can be considered for breast-cancer survivors if other treatments have not been useful, and can lead to a real improvement in genital (vulval, vaginal, and urinary) symptoms.

Non-hormonal treatment options include SSRI and SNRI antidepressants and anti-seizure medications such as pregabalin and gabapentin. This does not mean that your doctor thinks that you have depression or anxiety, rather that these medications can be used to help your symptoms. It is important to note that breast-cancer survivors should not use the antidepressants fluoxetine, paroxetine or duloxetine for treatment of menopausal symptoms or other conditions if taking tamoxifen. This is because for tamoxifen to work, it needs to be converted into its active form by an enzyme in the liver and these other medications may inhibit that process.

# breast changes, bras, and prostheses

Following breast-cancer treatment, you might need to buy new,
soft, comfortable bras and may be considering a prosthesis.

Your breasts may continue to change after treatment, so continue to be aware of how they look and feel, and report any changes you notice to your oncologist. Radiation therapy, for example, can change the breast, making it firmer and smaller than the other breast. Over time, as the breasts change after menopause to become softer, or you gain or lose weight, the breast which has had radiation therapy may not change as noticeably, so any asymmetry may become more apparent.

There may be ongoing reconstructive surgeries which affect the shape of the breasts. Cosmetic surgery can be carried out for asymmetry or to improve the appearance of indentations.

Breast-conserving surgery (a lumpectomy) aims to preserve the breast itself, yet you might still notice changes to the shape and size of the breast over time. There may be indents over the scar, if internal scar tissue tethers the skin.

## Wearing a bra

Initially after surgery, your breasts are likely to be tender, and the chest and armpit area may feel sensitive, so wear a soft, nonrestrictive bra for the first 6-8 weeks after surgery, or throughout radiation therapy. It may be difficult to put on a back-fastening bra due to restricted shoulder movement for the first few weeks, so some people prefer to fasten the clasp at the front and then move it around, or invest in a front-fastening bra.

In the first year after breast surgery it is recommended that you choose a bra with a wide, supportive band; soft seams without underwires; and full cups. This is because it can take up to a year to fully recover from surgery, and nerve stretching or injury may mean that you can't feel if an underwire is digging in.

Your bra size may change during and after breast-cancer treatment so be sure to check your bra fitting on multiple occasions. Post-mastectomy swimwear is also available.

## Wearing a prosthesis

If you have had a mastectomy, but not had breast reconstruction at the same time, you may choose to use a breast prosthesis in your bra. Initially these tend to be soft and made of fabric. If you choose not to have breast reconstruction, you may choose to utilize a permanent silicone prosthesis.

Bras are available with bra pockets which are designed to keep a breast prosthesis in place. Your local breast center may know someone who can sew a pocket into existing bras.

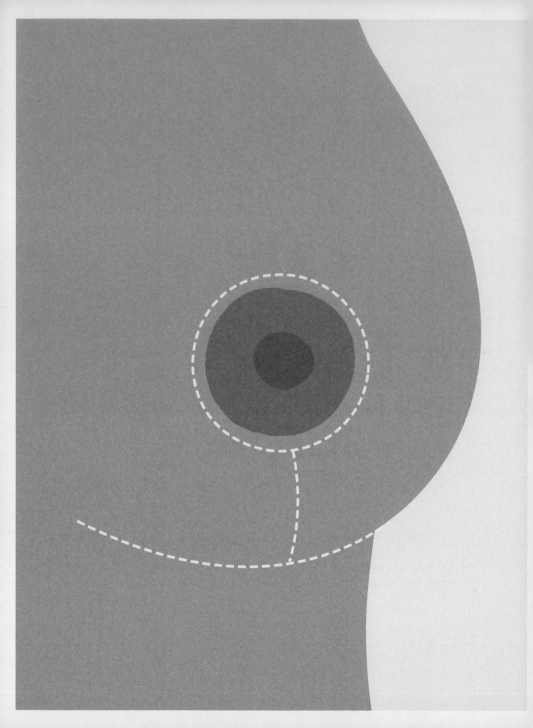

# 09

# Cosmetic surgery

# why have cosmetic surgery?

In this chapter we look at surgery to change the appearance of breasts, including reconstruction, reduction, and augmentation.

There are many reasons someone might consider breast surgery, such as to reduce the size of the breasts to relieve back and shoulder pain, correcting severe asymmetry, and reconstruction after breast-cancer surgery. Breast augmentation, aka "the boob job" is the most common cosmetic surgery performed globally, with nearly 1.8 million breast enlargement surgeries performed in 2019.

The psychological impact of breast surgery should not be underestimated. Breast reconstruction, if chosen, is considered part of breast-cancer treatment. Our identities are often interwoven with our appearances, and breast surgery can be life-changing for many, in terms of self-esteem. Perhaps your body has changed due to life-saving cancer surgery, from pregnancy or breastfeeding, or from weight changes, and surgery will help you feel more like your old self again. Breast surgery is also performed to meet the needs of trans and nonbinary people, with breast removal (bilateral mastectomy, or top surgery) and augmentation. There are lots of reasons, and there are also lots of types of surgeries.

## A brief history

Cosmetic surgery to change the appearance of breasts is not a new phenomenon. The first recorded breast-augmentation surgery was in 1889, when the surgeon used paraffin oil to enlarge the breast, but these early interventions had a high risk of infection. In 1895, a surgeon transplanted a lipoma (benign fatty mass) from one part of the patient's body into her chest, though over time the fat reabsorbed into the body.

Since the 19th century, all kinds of objects have been used as implants, including ivory, sponges, and glass, with high complication rates. In the mid-20th century, silicone was injected directly into the breasts, which sometimes led to gangrene. The first breast augmentation using silicone implants was performed in the US on Timmie Jean Lindsey in 1962.

## What is the cost?

Depending on where you live, it may be possible to obtain funding for particular types of surgery. For example, breast-reconstruction surgery may be covered. Coverage might also be offered for breast reduction for pain or psychological impact, or for asymmetry surgery or augmentation if you have a condition where breasts did not develop. However, in many cases the surgery is considered cosmetic and would need to be privately funded. Although the cost varies around the world, a breast augmentation performed privately currently costs £3,500-8,000 in the UK, and $5,000-10,000 in the US.

( . ) ( . )

# breast enlargement

Surgery to increase the size of the breasts, also known as augmentation, uses the insertion of implants. Let's look at what is involved, the recovery process, and potential safety concerns.

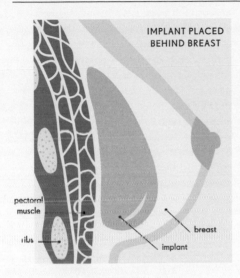

IMPLANT PLACED
BEHIND BREAST

pectoral
muscle

ribs

breast

implant

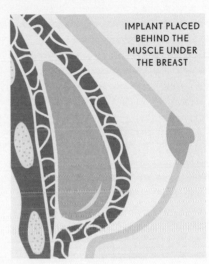

IMPLANT PLACED
BEHIND THE
MUSCLE UNDER
THE BREAST

Breast enlargement requires surgery under general anesthetic and involves the insertion of an implant. An implant can be placed underneath the breast tissue, or underneath the muscle, which itself is underneath the breast. Sometimes, the implant might be placed partly behind the breast tissue and partly under the muscle.

Depending on the surgery, the scars from the incisions may be in the inframammary fold—the line or fold under the breast—around the areola, or in the armpit.

Implants are generally made of a silicone elastomer "shell" and are filled with silicone or saline (salt water). Saline implants may feel harder and less "natural" than silicone. Implants can be of various sizes and shapes, including round or teardrop-shape. Teardrop-shape implants are sometimes called anatomical implants as they more closely replicate the shape of breasts, while round implants have more volume at the top of the breast. You and your surgeon will discuss the options together to find the best solution for your needs.

Topical creams sold that promise to enlarge your breasts don't work.

# How long do implants last?

Breast implants are not intended to last for your entire lifetime. Over time, you may require additional procedures; approximately 1 in 3 people will undergo a revision procedure. Most implants are approved for use for 10–20 years, but they do not have to be replaced after this point—only if they are causing problems or you are unhappy with their appearance.

Also bear in mind that if you gain or lose weight, your breasts will change shape and size and will still age, whether you have implants or not.

## RECOVERY AFTER SURGERY

Full recovery from breast-augmentation surgery takes at least six weeks, and it is advised you take at least two weeks off work. You may not be able to drive until you can safely do an emergency stop, which might be around two weeks after surgery. You may be advised to wear a supportive bra, both night and day, for up to three months. Your surgeon will give instructions regarding their post-surgery protocols.

## COMPLICATIONS OF SURGERY

Any surgery carries the risk of complications such as infection, bleeding, and blood clots. There are additional potential risks with breast-augmentation surgery. There may be a change in sensation to the breasts such as numbness, often around or near the scar, or the nipples may become extremely sensitive. These may improve in time but can also be permanent.

An implant is a foreign body, meaning that it isn't part of you, so your body will form a coating or capsule of scar tissue around it. For most people this will be so thin as to be unnoticeable, but in some people, a condition known as capsular contraction can occur. This is where the capsule of scar tissue becomes very thick, or contracts, and may cause pain or make the breast feel very hard. Newer designs of implant are less likely to cause this than older ones. About 1 in 10 breast-augmentation surgeries will have capsule contracture to some degree, and this may not appear for years after the surgery itself. If it is causing a problem then it may be possible to perform a capsulectomy (capsulotomy), where both the implant and capsule are removed.

Another complication is that it may be possible to see the fluid within the implant as it moves under the skin; this is more likely with saline than silicone implants. The implant may also move, or fold inside the breast, affecting the shape.

An extremely rare complication of breast-implant augmentation is breast implant associated large cell lymphoma (BIA-ALCL, see page 153) which is a form of blood cancer that affects the white blood cells (lymphoma) and presents with a swollen breast.

## Implant concerns

There is a small risk of rupture with all implants. If a saline implant ruptures it will not cause significant harm to you, as it contains salt water which will be absorbed by the body. However, the implant will go flat, affecting the appearance of the breast.

Silicone implants are made of a cohesive gel, meaning that the gel sticks together, so if there is a rupture it will not leak out, and the thin capsule of scar tissue around the implant keeps it in place. However, sometimes a ruptured silicone implant leads to small lumps which may be tender, or they may only be seen on scanning.

If you have large implants and then choose to have them removed for a smaller appearance later in life, the skin of the breasts may have been stretched, so the breast may look less full afterward.

You might have read about safety concerns regarding silicone implants, including lawsuits, risk of connective-tissue disorders, and cancer. Silicone implants were removed from use for a period of time, before the FDA (Food and Drug Administration) stated that they were safe for use in 2006.

## Poly implant prostheses (PIP) implants

PIP implants were withdrawn from use in the UK in 2010. The implants were found to have been made with an unapproved silicone gel which made them more prone to rupture than other silicone implants. It is estimated that approximately 47,000 women in the UK had PIP implants inserted. If you have had these implants, they do not have to be removed on medical grounds, as the research has not found that they are a significant health risk, although anxiety around their potential rupture can be a concern. If you have symptoms of rupture, such as swelling or lumpiness in the breast, change in skin color, or pain, please see your doctor. If it is found on scanning that the implant has ruptured, then you may be offered surgery to remove them.

---

### • AUGMENTATION AND BREAST CANCER •

There is no evidence that having breast augmentation increases the risk of developing breast cancer in the future. BIA-ALCL (see opposite) is a blood cancer, not a breast cancer, and extremely rare (1 in 30,000 patients, according to the FDA). For information about breast screening and implants, see page 124.

---

(.)(.)

# breast reduction and lift

Here, we look at the process involved in reducing breast size—often
performed to relieve pain and discomfort—and lifting the breast.

## Breast reduction

A breast reduction (reduction mammoplasty)
involves removing excess breast tissue, fat, and
skin, so the breast can be made into a smaller
size and shape. It is often performed to relieve
symptoms of pain (commonly neck, shoulder,
and back) caused by having large breasts, and
can also help with self-esteem and mental-
health issues related to having large breasts.
Reduction surgery tends to involve making
an incision under the fold of the breast
(inframammary fold), around the areola, and
then in a line vertically down to beneath the
breast. These can be shaped like a lollipop
(see top right) or an anchor (see bottom right).
The nipple may be moved to a higher position
in the breast.

   Do be aware that if there is a significant
amount of weight gain or loss later in your life,
or during pregnancy, then your breasts can
change size or shape again, even after
breast-reduction surgery.

### REDUCTION INCISIONS AND
POTENTIAL SCAR SITES

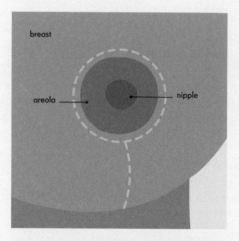

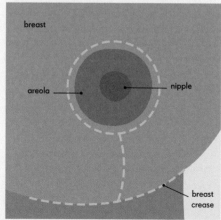

## Recovery after surgery

Recovery after a breast reduction generally takes about six weeks, and requires rest from work and not driving for a short period of time (see page 171 for more on recovery after breast surgery). It is likely that your surgeon will recommend that you wear a well-fitting, supportive bra for six to eight weeks after the surgery to help support your breasts while they are healing.

## Complications of surgery

In addition to the general complications of having any surgery, there is also a risk of loss of the nipple areolar complex, especially if it has been moved. You may also experience loss of sensation in the breast and nipple. There may not be enough breast tissue left for breastfeeding (see page 103).

## Breast-lift surgery

Surgery to lift the breasts (mastopexy) aims to elevate and reshape the breasts. The incision is made around the areola and then in a vertical line down to the fold under the breast. The excess skin is removed, the breast is reshaped, and the nipple may be repositioned higher on the breast. The areola can also be made smaller if required. In addition to the general complications of any surgery, there is also a risk of loss of the nipple and loss of sensation. Recovery generally takes around six weeks.

## Breast-symmetry procedures

Surgery may be performed to correct asymmetry which has been present since puberty, or to treat asymmetry which occurs later, perhaps after treatment for cancer. Some breast asymmetry is, however, completely normal. Surgery to correct breast asymmetry may involve breast reduction, breast augmentation, or a breast lift. It may be that after the breast-symmetry procedure each breast feels different; for example, if one contains an implant and the other does not.

### • MANAGEMENT OF SCARS •

Surgery leaves scars which take more than one year to fully remodel and heal. Initially, scars will look red and may be raised, but over time this is likely to fade. Scars can become widened or thickened, and some people may develop keloid scars, which are bigger than the original scar and can be red, itchy, and raised. Keloid scarring is more likely in melanated skin.

Once your surgeon has instructed it is safe to do so, regularly massaging your scars with a plain emollient can help reduce their appearance.

It is also important to use sunscreen to protect the skin from sun damage.

# breast reconstruction

Breast reconstruction is part of the treatment for breast cancer,
to recreate the shape and size of the breast.

## BREAST RECONSTRUCTION WITH AN IMPLANT

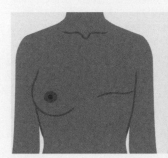 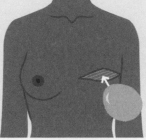 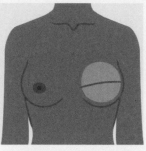

mastectomy scar (incision site)     implant inserted     implant in place under the skin

If you are having a mastectomy for breast
cancer, options for breast reconstruction are
discussed before the surgery to remove the
affected breast. These are big decisions made at
a time when you are dealing with a diagnosis of
cancer, with all that entails. Discuss your options
with your surgeon; support groups may also be
able to offer advice or peer support.

You may choose not to have reconstruction
at all, which is a valid and personal choice. You
might opt for immediate reconstruction, where
the reconstruction happens at the same time
as the mastectomy, or delayed reconstruction,
months or even years later, either because your
doctors have recommended waiting until after
chemotherapy, or because of personal choice.

Immediate reconstruction tends to give the
best results as the implant is put into your
breast right away. If you opt for a delayed
reconstruction, the skin of the breast has been
removed, so the surgeon has to either use skin
from a flap, or stretch your remaining skin with
an expander in order to create space for the flap
or implant. These options will result in differing
scars, with the scar from an immediate
reconstruction generally being around the areola
and a delayed reconstruction having a larger scar.

If you have large, pendulous breasts then a
reconstruction on one side with an implant is
unlikely to match the unaffected side unless a
breast reduction is carried out on the unaffected
side at the same time.

## Types of breast reconstruction

Generally there are two options for breast reconstruction: surgery involving an implant, or a flap-based reconstruction.

Implant surgery is often recommended for smaller breasts, where the implant is placed during a one- or two-stage operation. In a one-stage operation, a permanent implant is put in right away. In a two-stage operation, a temporary implant or expander is inserted first. A tissue expander is filled gradually with water over a period of weeks and months to slowly stretch the area. Once it has reached the right size, a second surgery is carried out to place a permanent implant. Or, an expander might be left in place, but the port through which it was filled is removed. A breast implant has a life span (see page 188) and additional surgeries may be needed in the future.

If you are having radiation therapy following your cancer surgery, your surgeon may suggest flap-based reconstruction to avoid implants. This is because radiation therapy can cause capsular contraction (see page 188).

Flap-based reconstruction does not use implants, or any other devices. It uses your own fat or muscle, which is removed from a donor area on your body (usually the tummy, bottom, or thigh), or is rotated from the back or abdomen, to create the breasts.

Flap-based reconstructions look, feel, and move like breasts. However, this is a more complex surgery, with a longer anticipated hospital stay as well as prolonged recovery periods (3-6 months). Depending on the flaps used there will be at least two scars, so there are increased risks of complications, such as wound infections. While a flap reconstruction doesn't have a life-span like an implant does, you may find that additional surgeries are needed to create symmetry.

## Bras and prostheses

Post-surgery bras are available, some of which are front-fastening to make them easier to put on. After a reconstruction, you will usually be advised to wear a bra, often night and day, for a few weeks, to support the breasts, which may help reduce discomfort and pain. It is important to look for adjustable straps and multiple fastenings so you can make the bra looser or tighter as required, as there is likely to be some post-operative swelling. Front-fastening bras may be easier to put on.

Fabric prostheses ("softies") are available for temporary use. They come in multiple sizes, shapes, and colors to match your skin tone. Some people like to stitch them into a bra so they don't move around.

Permanent breast prostheses are made from silicone and feel soft. They can be worn directly in the cup of your bra, or some bras have small pockets in the cups. Stick-on prostheses are also available. If you choose to wear a permanent breast prosthesis you can be fitted for one once you have fully healed from your surgery, and any swelling has resolved. This is generally 6-8 weeks after surgery. Prosthetic nipples are also available.

( . )( . )

# breast surgery

## SHOULD I HAVE BREAST RECONSTRUCTION AFTER MASTECTOMY?

This is a personal, individual choice; there is no right or wrong answer here. In the UK, of the people having a mastectomy each year, approximately 21 percent have immediate breast reconstruction and 10 percent have delayed reconstruction. The remaining 70 percent who have a mastectomy choose not to have reconstruction. If you are offered radiation therapy, some surgeons would advise not to have an implant because the radiation therapy can increase the risk of capsular contraction (see page 188). If you smoke, then your surgeon may advise against reconstruction because smoking can impact healing.

———

## CAN I KEEP MY NIPPLE IF I HAVE A MASTECTOMY?

It depends on the size of your breast and the site of the cancer. If the cancer is close to the nipple, or involves the nipple, the nipple may need to be removed. If the cancer is in another part of the breast, then a nipple-sparing mastectomy may be possible, though this very slightly increases the risk of recurrence of the cancer. If you have large breasts it may be difficult to save the nipple irrespective of where the cancer is, as it is harder to ensure that the nipple maintains its blood supply.

It is possible to have a nipple reconstruction. Your surgeon would use a flap of skin, and medical tattooing can be used to give the color and appearance of your areola and nipple. If you choose not to reconstruct the nipple, silicone prosthetic nipples are also available, if you would like.

———

## AM I TOO YOUNG FOR COSMETIC BREAST SURGERY?

With regard to breast reduction and asymmetry surgeries, it is generally advised to wait until the breasts have finished growing, generally after the age of 18. The FDA has approved silicone implants for those over 22 and saline for those over 18.

———

## WILL BREAST SURGERY IMPACT MY SEX LIFE AND LIBIDO?

Libido is extremely complicated and involves both physical and psychological factors. If the appearance of your breasts was negatively impacting your self-esteem and confidence, having breast surgery could well improve these symptoms, which may then impact your libido. However, the surgery itself may lead to nerve damage to the skin of the breasts and nipples, meaning that they become less sensitive, which may reduce sexual pleasure.

---

## WILL I NEED MORE SURGERY?

Whether or not you need more surgery will depend on a number of factors, including the type of surgery you have had. Breast augmentation is not a lifelong surgery, and around 10 percent of people will need further surgeries after a decade. Depending on the type of breast reconstruction surgery you have had, you may choose to have additional surgeries over time. For example, if you have an implant on one side only, the other breast may sag or change shape over time, and you may choose to have further surgery to improve symmetry.

---

## WILL I BE ABLE TO BREASTFEED AFTER SURGERY?

If you have had breast-reduction surgery there may not be sufficient breast tissue left for exclusive breastfeeding. Breast augmentation with implants that sit below the chest muscle are less likely to interfere with breastfeeding than those which sit above the muscle. Surgeries in which the nipple and areola are moved can lead to a reduced milk supply because the scar tissue may mean less milk can be delivered, and therefore less will be made.

---

# epilogue

For too long, women have been defined by their breasts; they have been sexualized, objectified, and vilified. Women have also been marginalized by science; most medical and scientific research has focused on men. Only in recent years has it become increasingly clear, in the words of Dr. Stacy Sims, that women are not merely small versions of men. Car crash dummies, stab vests, and even CPR mannequins were designed for those without breasts.

We know the importance of exercise to both physical and mental health, yet despite this, there is very little research into the impact of breast movement during exercise, even at elite levels. Young women are less likely to take part in regular exercise than men.

Bra sizes are not standardized across manufacturers, and the majority of women are wearing ill-fitting bras, which can mean pain and discomfort. We need the right bras for every shape and size, and at each stage of our lives, during exercise, after surgery, for breastfeeding, and beyond.

It is time to take control of the narrative. In order to take care of our bodies we have to get to know them. We have to be able to know our own normal, so that we can identify when something has changed. Normal for your breasts might be different to someone else's—after all, breasts can be big, small, and every size and shape in between. You are the best person to know your own body. And it isn't only women and girls who should be checking their breasts; everyone has some breast tissue, which means that everyone needs to check. Around 70 percent of us will get breast pain at some point so you need to know what is causing it, and when it might be a problem.

Breast cancer is the most common cancer (one in eight of us will get breast cancer) and we desperately need more research into the causes, treatments, and more. Most of the funding for breast health goes into breast cancer, yet it needs even more, in particular into advanced disease. But we also need more research into other areas of breast health. It isn't that I am calling for breast cancer to have less funding;

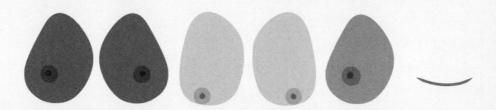

I am calling for breast cancer, breast health, and *all* other areas of women's health to have more.

The first point I want you to take from this book is to become "breast aware." Be aware of the general shape, size, and appearance of your breasts. Be on the lookout for any changes in the color and texture of your skin, any lumps, pain, or nipple discharge. If you do notice any changes, get it checked out. The second message I want to leave you with is: attend breast screening—it saves lives.

Breast health is not just about checking your breasts for potential issues; it is also about caring for your body so it can perform and function as you want it to. Studies show that young women want to learn more about their breasts and breast health, so we need to provide that information. Reading books like this are a start but we need to share this message—pass the book around. Ask at your kid's school if breast health is covered in the health or sexual education curriculum, and if not, challenge them to include it. Experts have compiled free resources and lesson plans available for teachers at www.treasureyourchest.org.

The internet is full of wonderful resources but it is also important that you know you are getting your information from a trusted source.

Susan G. Komen Breast Cancer Foundation (*www.komen.org*) is great for breast cancer advice; *www.knowyourlemons.org* and *coppafeel.org* have excellent guides to self-checks and breast awareness. There are also resources to help with the cost of screening if needed, such as the National Breast and Cervical Cancer Early Detection Program, NBCCEDP (*www.cdc.gov/cancer/nbccedp/*). Other sources may be less helpful—if an influencer is trying to sell you a product, then consider their content carefully; I've seen posts saying "bras cause breast cancer" which just isn't true. And keep talking—to your friends, your family, your partner—so the topics of breast health are no longer taboo and people know where to look for advice.

Now is the time for change, the time to own our bodies—including our breasts. Whatever your breasts look like, whatever you feel about them, whatever you want to do with them, they are yours. You have the right to call them their anatomical name—breasts aren't shameful or funny, to be hidden in slang and nicknames. We need to be literate about our bodies.

In order to own our breasts, we need to know them, take care of them, and then help others, too. Share this book, and spread the word!

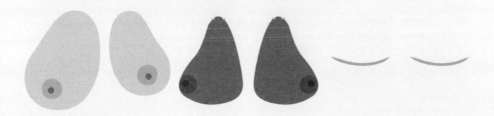

# bibliography

**Introduction**
**9** Scurr, J., Brown, N., Smith, J., Brasher, A., Risius, D., & Marczyk, A. (2016). The influence of the breast on sport and exercise participation in schoolgirls in the UK. Journal of Adolescent Health. 58(2), 167-173 · www.who.int/about/governance/constitution · www.wcrf.org/cancer-trends/worldwide-cancer-data/

**Chapter 1 / Breasts are for ...**
**13** www.sciencedirect.com/science/article/abs/pii/S1090513816302847 · Viren Swami, Martin J. Tovée, Resource Security Impacts Men's Female Breast Size Preferences, 2013 doi.org/10.1371/journal.pone.0057623 · Dixson BJ, Vasey PL, Sagata K, Sibanda N, Linklater WL, Dixson AF. Men's preferences for women's breast morphology in New Zealand, Samoa, and Papua New Guinea. Arch Sex Behav. 2011 Dec;40(6):1271-9. doi: 10.1007/s10508-010-9680-6. · Zelazniewicz, A.M., Pawlowski, B. Female Breast Size Attractiveness for Men as a Function of Sociosexual Orientation (Restricted vs. Unrestricted). Arch Sex Behav 40, 1129-1135 (2011). doi.org/10.1007/s10508-011-9850-1 · Patterns of Sexual Behavior, C. S. Ford, F. A. Beach, (1951), Psychology · **14** ZAVA (online doctor provider): adapted from: https://www.zavamed.com/uk/does-breast-size-matter.html. **15** Evolution and Human Behaviour, Volume 38, Issue 2, March 2017, Pages 217-226 · **19** www.asa.org.uk/rulings/adidas-uk-ltd-g22-1145614-adidas-uk-ltd.html

**Chapter 2 / Anatomy and everyday maintenance**
**28** StatPearls Treasure Island (FL): StatPearls Publishing; 2022 Jan. · Anatomy, Shoulder and Upper Limb, Axillary Lymph Nodes, Harry Kyriacou; Yusuf S. Khan. www.ncbi.nlm.nih.gov/books/NBK430685/ **29** BMJ. 2008 Mar 29; 336(7646): 709-713. doi: 10.1136/bmj.39511.493391.BE · Leung AKC, Leung AAC. Gynecomastia in Infants, Children, and Adolescents. Recent Pat Endocr Metab Immune Drug Discov. 2017;10(2):127-137. doi: 10.2174/1872214811666170301124033. **30** pedsendo.org/patient-resource/premature-thelarche/ **31** www.healthline.com/health/breast-asymmetry **32** Sanuki J ichi, Fukuma E, Uchida Y. Morphologic study of nipple-areola complex in 600 breasts. Aesth Plast Surg. 2009;33(3):295-297. doi: 10.1007/s00266=088-9194-y. **34** Zucca-Matthes G, Urban C, Vallejo A. Anatomy of the nipple and breast ducts. Gland Surg. 2016;5(1):32-6. doi:10.3978/j.issn.2227-684X.2015.05.10 **37** Nagaraja Rao D, Winters R. Inverted nipple. StatPearls. **38** www.breastcancer.org/facts-statistics · www.cancerresearchuk.org/health-professional/cancer-statistics/statistics-by-cancer-type/breast-cancer **39** cewuk.co.uk/estee-lauder-companies-marks-30th-anniversary-of-breast-cancer-campaign **40-43** www.nhs.uk/common-health-questions/womens-health/how-should-i-check-my-breasts/ **43** ascopost.com/issues/january-25-2019/risk-of-local-recurrence-in-breast-cancer/ · breastcancernow.org/information-support/check-your-breasts/learn-signs-breast-cancer

**Chapter 3 / During puberty**
**46** www.nhs.uk/conditions/early-or-delayed-puberty/ **47** www.ncbi.nlm.nih.gov/books/NBK470280/ **48-49** Caouette-Laberge L, Borsuk D. Congenital anomalies of the breast. Semin Plast Surg. 2013 Feb;27(1):36-41. doi: 10.1055/s-0033-1343995. · www.bapras.org.uk/public/patient-information/surgery-guides/congenital-breast-and-chest-conditions **49** Wolfswinkel EM, Lemaine V, Weathers WM, Chike-Obi CJ, Xue AS, Heller L. Hyperplastic breast anomalies in the female adolescent breast. Semin Plast Surg. 2013 Feb;27(1):49-55. doi: 10.1055/s-0033-1347167. **52** McGhee DE, Steele JR. Optimising breast support in female patients through correct bra fit. A cross-sectional study. J Sci Med Sport. 2010 Nov;13(6):568-72. doi: 10.1016/j.jsams.2010.03.003. **61** White, J., & Scurr, J. (2012). Evaluation of professional bra fitting criteria for bra selection and fitting in the UK. Ergonomics, 55(6), 704-711. doi.org/10.1080/00140139.2011.647096

**Chapter 4 / After puberty**
**74** Coltman CE, Steele JR, McGhee DE. Does breast size affect how women participate in physical activity? J Sci Med Sport. 2019 Mar;22(3):324-329. doi: 10.1016/j.jsams.2018.09.226. · Burnett E, White J, Scurr J. The Influence of the Breast on Physical Activity Participation in Females. J Phys Act Health. 2015 Apr;12(4):588-94. doi: 10.1123/jpah.2013-0236. · Gehlsen G, Albohm

M. Evaluation of Sports Bras. Phys Sportsmed. 1980 Oct;8(10):88-97. doi: 10.1080/00913847.1980.11948653. **75** Sports Medicine Australia: adapted from: https://sma.org.au/resources-advice/injury-fact-sheets/exercise-and-breast-support/ · Mason BR, Page KA, Fallon K. An analysis of movement and discomfort of the female breast during exercise and the effects of breast support in three cases. J Sci Med Sport. 1999 Jun;2(2):134-44. doi: 10.1016/s1440-2440(99)80193-5. · Scurr JC, White JL, Hedger W. Supported and unsupported breast displacement in three dimensions across treadmill activity levels. J Sports Sci. 2011 Jan;29(1):55-61. doi: 10.1080/02640414.2010.521944. · Bridgman C, Scurr J, White J, Hedger W, Galbraith H. Three-dimensional kinematics of the breast during a two-step star jump. J Appl Biomech. 2010 Nov;26(4):465-72. doi: 10.1123/jab.26.4.465. **76** Scurr J, Brown N, Smith J, Brasher A, Risius D, Marczyk A. The Influence of the Breast on Sport and Exercise Participation in School Girls in the United Kingdom. J Adolesc Health. 2016 Feb;58(2):167-73. doi: 10.1016/j.jadohealth.2015.10.005. · womenandsport.ca/resources/research-insights/rally-report/ · www.bbc.co.uk/news/health-60646352 · Scurr J, Brown N, Smith J, Brasher A, Risius D, Marczyk A. The Influence of the Breast on Sport and Exercise Participation in School Girls in the United Kingdom. J Adolesc Health. 2016 Feb;58(2):167-73. doi: 10.1016/j.jadohealth.2015.10.005. **77** McGhee DE, Steele JR. Biomechanics of Breast Support for Active Women. Exercise and sport sciences reviews. 2020 Jul 1;48(3):99-109. pubmed.ncbi.nlm.nih.gov/32271181/ · McGhee DE, Steele JR. Breast elevation and compression decrease exercise-induced breast discomfort. Med Sci Sports Exerc. 2010 Jul;42(7):1333-8. doi: 10.1249/MSS.0b013e3181ca7fd8. **80** Coltman, C.E., McGhee, D.E. & Steele, J.R. Bra strap orientations and designs to minimise bra strap discomfort and pressure during sport and exercise in women with large breasts. Sports Med—Open 1, 21 (2015). doi.org/10.1186/s40798-015-0014-z **81** A. Dhivya, V. P. R. P. ,S. S. (2016). Design and Development of Sports Intimate Apparel—A Review. SMART MOVES JOURNAL IJOSTHE, 3(1). ijosthe.com/index.php/ojssports/article/view/52 · Norris M, Blackmore T, Horler B, Wakefield-Scurr J. How the characteristics of sports bras affect their performance. Ergonomics. 2021 Mar;64(3):410-425. doi: 10.1080/00140139.2020.1829090. · Joanna Wakefield-Scurr et al. The effect of washing and wearing on sports bra function. Sports Biomechanics, 2022 DOI: 10.1080/14763141.2022.2046147 **82** Peitzmeier S, Gardner I, Weinand J, Corbet A, Acevedo K. Health impact of chest binding among transgender adults: a community-engaged, cross-sectional study. Cult Health Sex. 2017 Jan;19(1):64-75. doi: 10.1080/13691058.2016.1191675. · www.prideinpractice.org/articles/chest-binding-physician-guide · Brooke A. Jarrett, et al. Chest Binding and Care Seeking Among Transmasculine Adults. Transgender Health. Dec.www.ncbi.nlm.nih.gov/pmc/articles/PMC6298447/ **83** patient.info/news-and-features/how-to-bind-your-chest-safely **84** www.healthline.com/health/body-modification/nipple-piercing-aftercare **85** www.painfulpleasures.com/community/blog/client/nipple-piercing-guide-everything-you-need-to-know/ · www.medicalnewstoday.com/articles/318148

## Chapter 5 / Pregnancy and breastfeeding

**88-97** Motosko CC, Bieber AK, Pomeranz MK, Stein JA, Martires KJ. Physiologic changes of pregnancy: A review of the literature. Int J Womens Dermatol. 2017 Oct 21;3(4):219-224. doi: 10.1016/j.ijwd.2017.09.003. · www.hopkinsmedicine.org/health/conditions-and-diseases/normal-breast-development-and-changes · www.healthline.com/health/pregnancy/pregnant-breast · breastcancernow.org/information-support/have-i-got-breast-cancer/breast-changes-during-after-pregnancy · sthk.nhs.uk/media/.leaflets/G11a01f02790b0.27507630.pdf **90** Geraghty LN, Pomeranz MK. Physiologic changes and dermatoses of pregnancy. Int J Dermatol. 2011 Jul;50(7):771-82. doi: 10.1111/j.1365-4632.2010.04869.x. **93** www.who.int/health-topics/breastfeeding **95** Jaclyn Pillay & Tammy J. Davis. Physiology, Lactation. National Library of Medicine. July 2022 **98** www.forbes.com/health/family/unable-to-breastfeed/ **100-101** www.nhs.uk/conditions/baby/breastfeeding-and-bottle-feeding/breastfeeding/positioning-and-attachment/ **106** Boi B, Koh S, Gail D. The effectiveness of cabbage leaf application (treatment) on pain and hardness in breast engorgement and its effect on the duration of breastfeeding. JBI Libr Syst Rev. 2012;10(20):1185-1213. doi: 10.11124/01938924-201210200-00001 **110** Wilson E, Woodd SL, Benova L; Incidence and Risk Factors for Lactational Mastitis: A Systematic Review. J Hum Lact. 2020 Nov36(4):673-686. doi: 10.1177/0890334420907898. **111** Lisa H. Amir, et al. Volume 111 Issue 12 Incidence of breast abscess in lactating women: report from an Australian cohort. International Journal of Obstetrics and Gynaecology. Nov 2004, doi.org/10.1111/j.1471-0528.2004.00272.x

## Chapter 6 / Menopause and beyond

**117** Martinez AA, & Chung S. Breast Ptosis. National Library of Medicine. Jan 2022 **119** www.cdc.gov/cancer/breast/pdf/breast-cancer-screening-guidelines · digital.nhs.uk/data-and-information/publications/statistical/breast-screening-programme/england---2019-20 · Independent UK Panel on Breast Cancer Screening. The benefits and harms of breast cancer screening: an independent review. Lancet. 2012 Nov 17;380(9855):1778-86. doi: 10.1016/S0140-6736(12)61611-0. · Hendrick RE, Baker JA, Helvie MA. Breast cancer deaths averted over 3 decades. Cancer. 2019 May 1;125(9):1482-1488. doi: 10.1002/cncr.31954. · Gøtzsche PC, Jørgensen KJ. Screening for breast cancer with mammography. Cochrane Database Syst Rev. Jun 2013 **120** paho.org/hq/dmdocuments/2015/WHO-ENG-Mammography-Factsheet.pdf **121** digital.nhs.uk/data-and-information/publications/statistical/breast-screening-programme/england---2019-20 · www.urmc.rochester.edu/news/publications/health-matters/mammograms-facts-on-false-positives **123** www.gov.uk/government/publications/breast-screening-helping-women-decide **124** progressreport.cancer.gov/detection/breast_cancer **125** Bond M, Pavey T, Welch K, Cooper C, Garside R, Dean S, Hyde C. Systematic review of the psychological consequences of false-positive screening mammograms. Health Technol Assess. 2013 Mar;17(13):1-170, v-vi. doi: 10.3310/hta17130 · Waller J, Douglas E, Whitaker KL, Wardle J. Women's responses to information about overdiagnosis in the UK breast cancer screening programme: a qualitative study. BMJ Open. 2013 Apr 22;3(4):e002703. doi: 10.1136/bmjopen-2013-002703.

### Chapter 7 / When things go wrong
**128** Goyal A. Breast pain. The British Medical Journal, Clin Evid. Oct 2014 Goyal A. Breast pain. BMJ Clin Evid. 2014 Oct 14;2014:0812. PMCID: PMC4200534. · cks.nice.org.uk/topics/breast-pain-cyclical Richard Sadovsky, Topical NSAIDs Relieve the Pain of Mastalgia, American Family Physician. 2003 **129** Vaziri F, et al. Comparing the effects of dietary flaxseed and omega-3 Fatty acids supplement on cyclical mastalgia in Iranian women: a randomized clinical trial. International Journal of Family Medicine. Aug 2014 doi: 10.1155/2014/174532. · Goyal A. Breast pain. BMJ Clin Evid. 2014 Oct 14;2014:0812. PMCID: PMC4200534. **136** www.cancerresearchuk.org/cancer-symptoms/what-is-an-urgent-referral **137** www.gov.uk/government/publications/breast-screening-helping-women-decide/nhs-breast-screening-helping-you-decide **139** Malherbe K, Khan M, Fatima S. Fibrocystic Breast Disease. National Library of Medicine. Oct 2021 · Ajmal M, Khan M, Van Fossen K. Breast Fibroadenoma. Breast Fibroadenoma. National Library of Medicine. Apr 2022.

### Chapter 8 / Breast cancer
**145** Apostolou P, Fostira F. Hereditary breast cancer: the era of new susceptibility genes. Biomed Res Int. 2013;2013:747318. doi: 10.1155/2013/747318. · www.nhs.uk/conditions/predictive-genetic-tests-cancer · www.bccp.org/resource/african-american-women-and-breast-cancerCragun D, et al. Racial disparities in BRCA testing and cancer risk management across a population-based sample of young breast cancer survivors. Cancer. July 2017. · breastcancernow.org/about-us/media/facts-statistics/how-are-people-ethnically-diverse-backgrounds-impacted-breast-cancer · Jones M et al. Smoking and risk of breast cancer in the Generations Study cohort. Breast cancer research. Nov 2017. · www.webmd.com/breast-cancer/overview-risks-breast-cancer **146** www.komen.org/breast-cancer/screening/screening-disparities · Yedjou CG, Sims JN, Miele L, Noubissi F, Lowe L, Fonseca DD, Alo RA, Payton M, Tchounwou PB. Health and Racial Disparity in Breast Cancer. Adv Exp Med Biol. 2019;1152:31-49. doi: 10.1007/978-3-030-20301-6_3. · news.cancerresearchuk.org/2016/11/16/black-african-women-almost-twice-as-likely-to-be-diagnosed-with-late-stage-breast-cancer-compared-to/ **146** Suther, S., Kiros, GE. Barriers to the use of genetic testing: A study of racial and ethnic disparities. Genet Med 11, 655-662 (2009). doi.org/10.1097/GIM.0b013e3181ab22aa **148** Gonçalves AK et al. Effects of physical activity on breast cancer prevention: a systematic review. Journal of Physical Activity and Health. Feb 2014 · Kotepui M. Diet and risk of breast cancer. Contemp Oncol (Pozn). 2016;20(1):13-9. doi: 10.5114/wo.2014.40560. · www.nhs.uk/live-well/exercise/exercise-guidelines/physical-activity-guidelines-for-adults-aged-19-to-64 · www.breastcanceruk.org.uk/reduce-your-risk/physical-activity-and-exercise · Donaldson MS. Nutrition and cancer: a review of the evidence for an anti-cancer diet. Nutr J. 2004 Oct 20;3:19. doi: 10.1186/1475-2891-3-19. · J Connor, Alcohol consumption as a cause of cancer, Addiction, Feb 2017, doi.org/10.1111/add.13477 · breastcancernow.org/information-support/have-i-got-breast-cancer/breast-cancer-causes/alcohol-breast-cancer-risk · www.nhs.uk/live-well/alcohol-advice/calculating-alcohol-units/ **149** news.cancerresearchuk.org/2017/05/25/alcohol-and-breast-cancer-how-big-is-the-risk/ · Hamajima N et al. Collaborative Group on Hormonal Factors in Breast Cancer. · Alcohol, tobacco and breast cancer. British Journal of Cancer. 2002 Nov · World Cancer Research Fund/American Institute for Cancer Research. Continuous Update Project Findings & Reports. June 2017. **150** Thebms.org.uk/wp-content/uploads/2020/12/12-BMS-Tfc-Fast-Facts-HRT-and-Breast-Cancer-Risk-01D.pdf **152** www.fda.gov/food/food-additives-petitions/additional-information-about-high-intensity-sweeteners-permitted-use-food-united-states **153** Boutas I, et al. Soy Isoflavones and Breast Cancer Risk: A Meta-analysis. In vivo. Mar 2022. PMID: 35241506 · Finkeldey L, et al. Effect of the Intake of Isoflavones on Risk Factors of Breast Cancer. Nutrients. July 2021. **155** www.cancer.gov/about-cancer/causes-

prevention/genetics/brca-fact-sheet **156** Warner E et al. Prevalence and penetrance of BRCA1 and BRCA2 gene mutations in unselected Ashkenazi Jewish women with breast cancer. Journal of the National Cancer Institute. July 1999 doi: 10.1093/jnci/91.14.1241. **158** www.breastcanceruk.org.uk/breast-cancer-in-men/ **161** Office for National Statistics, Cancer survival by stage at diagnosis for England, 2019. **164** www.cancerresearchuk.org/about-cancer/breast-cancer/getting-diagnosed/tests-diagnose/hormone-receptor-testing-breast-cancer **165** Onitilo AA, Engel JM, Stankowski RV, Doi SA. Survival Comparisons for Breast Conserving Surgery and Mastectomy Revisited: Community Experience and the Role of Radiation Therapy. Clin Med Res. 2015 Jun;13(2):65-73. doi: 10.3121/cmr.2014.1245. **170** Jennifer A. et al, Exercise, Diet, and Weight Management During Cancer Treatment: ASCO Guideline, Journal of Clinical Oncology 40, no. 22 (August 01, 2022), DOI: 10.1200/JCO.22.00687 · Rikki A Cannioto et. al, Physical Activity Before, During, and After Chemotherapy for High-Risk Breast Cancer: Relationships With Survival, JNCI: Journal of the National Cancer Institute, Volume 113, Issue 1, January 2021, Pages 54–63, doi.org/10.1093/jnci/djaa046 · Holmes MD, Chen WY, Feskanich D, Kroenke CH, Colditz GA. Physical activity and survival after breast cancer diagnosis. JAMA. 2005 May 25;293(20):2479-86. doi: 10.1001/jama.293.20.2479. **171** breastcancernow.org/sites/default/files/publications/pdf/bcc6_exercises_booklet_2019_web.pdf **177** Fleissig A, Fallowfield LJ, Langridge CI, Johnson L, Newcombe RG, Dixon JM, Kissin M, Mansel RE. Post-operative arm morbidity and quality of life. Breast Cancer Res Treat. 2006 Feb;95(3):279-93. doi: 10.1007/s10549-005-9025-7. · DiSipio T, Rye S, Newman B, Hayes S. Incidence of unilateral arm lymphoedema after breast cancer: a systematic review and meta-analysis. Lancet Oncol. 2013 May;14(6):500-15. doi: 10.1016/S1470-2045(13)70076-7. **178** Couceiro TC, Valença MM, Raposo MC, Orange FA, Amorim MM. Prevalence of post-mastectomy pain syndrome and associated risk factors: a cross-sectional cohort study. Pain Manag Nurs. 2014 Dec;15(4):731-7. doi: 10.1016/j.pmn.2013.07.011. · Cui L, Fan P, Qiu Y, Hong Y. Single institution analysis of incidence and risk factors for post-mastectomy pain syndrome. Sci Rep. 2018 Jul 31;8(1):11494. doi: 10.1038/s41598-018-29946-x. · Beyaz SG et al. Postmastectomy Pain: A Cross-sectional Study of Prevalence, Pain Characteristics, and Effects on Quality of Life. Chin Med J (Engl). 2016 Jan 5;129(1):66-71. doi: 10.4103/0366-6999.172589. **179** www.komen.org/breast-cancer/treatment/recurrence/survival-and-risk-of-recurrence/ · Braunstein LZ et al., Breast-cancer subtype, age, and lymph node status as predictors of local recurrence following breast-conserving therapy. Breast Cancer Res Treat. 2017 Jan;161(1):173-179. doi: 10.1007/s10549-016-4031-5. · Arvold ND et al., breast cancer subtype approximation, and local recurrence after breast-conserving therapy. J Clin Oncol. 2011 Oct 10;29(29):3885-91. doi: 10.1200/JCO.2011.36.1105. **181** www.cancer.org/content/dam/cancer-org/research/cancer-facts-and-statistics/breast-cancer-facts-and-figures-2019-2020.pdf · www.cancer.org/cancer/breast-cancer/about/how-common-is-breast-cancer.html · Ahlberg K, Ekman T, Gaston-Johansson F, Mock V. Assessment and management of cancer-related fatigue in adults. Lancet. 2003 Aug 23;362(9384):640-50. doi: 10.1016/S0140-6736(03)14186-4. · Bower JE, Ganz PA, Desmond KA, Rowland JH, Meyerowitz BE, Belin TR. Fatigue in breast cancer survivors: occurrence, correlates, and impact on quality of life. J Clin Oncol. 2000 Feb,18(4).743-53. doi: 10.1200/JCO.2000.18.4.743. · Maass SWMC et al. Fatigue among Long-Term Breast Cancer Survivors: A Controlled Cross-Sectional Study. Cancers (Basel). 2021 Mar 15;13(6):1301. doi: 10.3390/cancers13061301.

### Chapter 9 / Cosmetic surgery
**186** www.statista.com/chart/25322/plastic-surgery-procedures-by-type/ · www.nhs.uk/conditions/cosmetic-procedures/breast-enlargement/ · www.plasticsurgery.org/cosmetic-procedures/breast-augmentation/cost · www.webmd.com/beauty/cosmetic-procedures-breast-augmentation **188** www.bapras.org.uk/docs/default-source/Patient-Information-Booklets/rcs_bapras_guide_breast_augmentation.pdf · Headon H et.al. Capsular Contracture after Breast Augmentation: An Update for Clinical Practice. Arch Plast Surg. 2015 Sep;42(5):532-43. doi: 10.5999/aps.2015.42.5.532. **189** www.nhs.uk/conditions/pip-implants/ · www.mskcc.org/news/fda-s-new-guidance-breast-implants-what-breast-cancer-patients-need-know · files.digital.nhs.uk/publicationimport/pub02xxx/pub02731/clin-audi-supp-prog-mast-brea-reco-2011-rep1.pdf

### Epilogue
**196** www.theguardian.com/lifeandstyle/2019/feb/23/truth-world-built-for-men-car-crashes · Gehlsen G, & Albohm M. Evaluation of sports bras. The Physician and Sports Medicine. Oct 1980 pubmed.ncbi.nlm.nih.gov/29261415/

To access a comprehensive list of source materials, studies and research supporting the information in this book, please visit: **www.dk.com/breasts-biblio**

( . ) ( . )

# index

# acknowledgments

## Author's acknowledgments

To my agent Jane Graham Maw of Graham Maw Christie and Jennifer Christie for being a constant source of advice and reassurance for what now is beginning to be too many years to be mentioned!

To Zara Anvari, senior acquisitions editor at Dorling Kindersley for being persistent in bringing this book to life. Lauren Mitchell and Hannah Naughton for translating my words into illustrations (I know what I want I just can't draw it!) and for the design by Hannah Naughton and Tania Da Silva Gomes. Thanks must also be given to editors Lucy Sienkowska and Becky Alexander for their patience, wisdom and input all done at breakneck speed. This book will have been touched by many more hands during its conception and delivery at Dorling Kindersley, from Sales to PR to proofreaders, I thank you all.

My colleagues in science and medicine for allowing me to interview them and for peer-reviewing sections of the book. Ms Liz O'Riordan, consultant breast surgeon and breast cancer survivor whose knowledge is unparalleled. Professor Joanna Wakefield-Scurr and the Research Group in Breast Health at the University of Portsmouth for her immense knowledge about the importance of, yet frustrating, sizing of bras and sports bras and all her and her team's work to try and improve the process. To Rebecca Sellars, breast cancer physiotherapist, for her advice on exercises and cording.

To my patients whom over the years have taught and continue to teach me about health, women's health, and breast health.

To my family, in particular my mother who taught me from an early age the importance of a good bra and has joined me in many bra shopping trips. In our house growing up, breasts were named wobbles by my older sister, as they, well, "wobbled"! To my husband Ben and my children, who put up with me writing at all times, including family film night (sorry, I really tried not to but the deadline was so, so tight!) and support me in every way possible. My children now know perhaps a surprising amount about the history of breasts and bras and walk into any museum or art gallery already on guard looking for representations of breasts!

To my friends, Vicki, Lucie, Kate and Susannah who held me against their metaphorical bosoms and supported me while I wrote this book.

And finally to you, for reading, and hopefully sharing the messages in this book. In the (sort of) words of Shakira, whether your breasts be small and humble, or confused with mountains, they are yours—we need to look after them.

## About the author

Dr. Philippa Kaye is a GP, presenter, author, and mother of three. She studied medicine at Downing College, Cambridge, followed by Guy's, King's, and St Thomas' medical schools. She completed hospital posts in various specialties across London before qualifying as a GP. She is a media doctor, appearing regularly on programmes such as *This Morning*, and *Vanessa Feltz on Talk TV* as well as various other television and radio programmes. She is the medical expert for magazines *Woman*, *Woman & Home*, *That's Life*, and *My Weekly Special*, and the *MadeforMums* website. Dr. Kaye has written many books, including *The M Word: Everything You Need to Know About the Menopause*, and *Doctors Get Cancer Too*, in which she shares her own story of being diagnosed with bowel cancer at the age of 39.

## Publisher's acknowledgments

DK would like to thank Karen L Wang, MD, for consulting; Ms Liz O'Riordan for peer reviewing; Alex Whittleton for proofreading; Ruth Ellis for the index; and Myriam Megharbi for data permissions clearing.

( . ) ( . )

**Project Editor** Becky Alexander
**Designer** Hannah Naughton
**Illustration** Hannah Naughton and Lauren Mitchell

DK

**Senior Acquisitions Editor** Zara Anvari
**Editor** Lucy Sienkowska
**US Editor** Megan Douglass
**Senior Designer** Tania Gomes
**Jacket Coordinator** Jasmin Lennie
**Senior Production Editor** Tony Phipps
**Production Controller** Rebecca Parton
**DTP and Design Coordinator** Heather Blagden
**Editorial Manager** Ruth O'Rourke
**Design Manager** Marianne Markham
**Art Director** Maxine Pedliham
**Publishing Director** Katie Cowan

First American Edition, 2023
Published in the United States by DK Publishing
1745 Broadway, 20th Floor, New York, NY 10019

Published in Great Britain by Dorling Kindersley Limited

A catalog record for this book
is available from the Library of Congress.
ISBN: 978-0-7440-7938-8

Printed and bound in Slovakia

**For the curious**
www.dk.com

## DISCLAIMER

The information in this book has been compiled by way
of general guidance in relation to the specific subjects
addressed. It is not a substitute and not to be relied on for
medical, health care, pharmaceutical, or other professional
advice on specific circumstances and in specific locations.
Please consult your doctor before starting, changing, or
stopping any medical treatment. So far as the author is
aware, the information given is correct and up to date as of
November 2022. Practice, laws, and regulations all change,
and the reader should obtain up-to-date professional
advice on any such issues. The naming of any product,
treatment, or organization in this book does not imply
endorsement by the author or publisher, nor does the
omission of any such names indicate disapproval. Neither
the author nor the publisher shall be liable or responsible
for any loss or damage allegedly arising from any
information or suggestion in this book.

## A NOTE ON GENDER IDENTITIES

DK recognizes all gender identities, and acknowledges
that the sex someone was assigned at birth based on their
sexual organs may not align with their own gender identity.
People may self-identify as any gender or no gender
(including, but not limited to, that of a cis or trans woman,
of a cis or trans man, or of a nonbinary person). As gender
language, and its use in our society, evolves, the scientific
and medical communities continue to reassess their own
phrasing. Most of the studies referred to in this book use
"women" to describe people whose sex was assigned as
female at birth, and "men" to describe people whose sex
was assigned as male at birth.